Social Media, Parties, and
Political Inequalities

Social Media, Parties, and Political Inequalities

Kristof Jacobs and Niels Spierings

First published 2016 by
PALGRAVE MACMILLAN

The authors have asserted their rights to be identified as the authors of this work in accordance with the Copyright, Designs and Patents Act 1988.

Palgrave Macmillan in the UK is an imprint of Macmillan Publishers Limited, registered in England, company number 785998, of Houndmills, Basingstoke, Hampshire, RG21 6XS.

Palgrave Macmillan in the US is a division of Nature America, Inc., One New York Plaza, Suite 4500, New York, NY 10004-1562.

Palgrave Macmillan is the global academic imprint of the above companies and has companies and representatives throughout the world.

ISBN: 978-1-349-57271-7
E-PDF ISBN: 978–1–137–53390–6
DOI: 10.1057/9781137533906

Distribution in the UK, Europe and the rest of the world is by Palgrave Macmillan®, a division of Macmillan Publishers Limited, registered in England, company number 785998, of Houndmills, Basingstoke, Hampshire RG21 6XS.

Library of Congress Cataloging-in-Publication Data is available from the Library of Congress.

A catalogue record for the book is available from the British Library.

Contents

Illustrations

Figures

Tables

Boxes

Acknowledgments

This book is the result of five years of research on the political impact of social media in the VIRAL-project. In 2009 it was not certain whether social media would break through or not, but our gamble—in far as one can speak about a gamble when studying something that simply interests oneself—to start investigating this new way of communicating paid off and resulted in several studies and now in this book where we bring together our ideas, further crystalized them, and have the opportunity to tell the larger story.

Writing this book would not have been possible without the help and support of many people. Our research interns Ploni Stoop, Willem van Sermondt, and Eirin Kofoed collected data on local politicians or carried out analysis exploring the professionalization of social media accounts. Jeroen Hellebrekers, Sophie Lauwers, Sjors Talsma, and Malu Verkuil, in the position of student-assistants, also helped us with data collection and entry as well as discussing ideas. Nik Linders created the "Superviral" website and his technical brilliance, and his help as a volunteer (!) were indispensible in collecting the data. The Department of Political Science supported the data collection financially and made this project possible. Our respective heads of departments believed in the project, agreed with us spending valuable research time on it (also when social media were not booming yet), and helped us make it possible (even though we sometimes handed in requests they found a bit strange, like our own server to store data). We also want to thank the people at Palgrave. Brian O'Connor scouted us, and later Nicole Hitner, Elaine Fan, and Alexandra Dauler provided invaluable support during and at the end of the writing process. The theoretical framework applied in the book benefited greatly from discussion with academic colleagues. In particular we benefited from the conversations with Karen Celis, Sanne Kruikemeier, Rens Vliegenthart, Laura Sudulich, the participants of the 2012 ECPR Joint Sessions workshop in Antwerp and the 2015 ECPR workshop in Warsaw.

Even though all these professional sources of inspiration were important, we want to end with some personal acknowledgements.

Kristof

I want to thank three important and inspiring women, namely Florentine Verhaert-Leunis, Emma Verhaert, and Meike Kool. Meike, you always know how to make me worry less and enjoy life even when deadlines draw near. I also want to thank my father and brother. Vinze, we did not use Python here, but I did use R. Wietse; thank you for being such an inspiration and...thanks for showing me how you use Facebook professionally in your work.

Niels

This book started as a "hobby project": something we were interested in and started to study in addition to our regular research. Consequently, we conducted much of the research in "hobby time," with, as consequence, less time for people who are very dear to me. In the last few months or writing particularly Ann, Mark, Monique, Ralf, Rieke, and Rogier suffered I think. In addition, thank you to all the people who keep following me on Facebook, Instagram, and Twitter despite the regular complaining about the above and other things.

Kristof Jacobs—T: @KristofJacobs1—L: kristofjacobs
Niels Spierings—T: @NielsSpierings—F: niels.spierings.9—L: nielsspierings

Abbreviations

CDA	ChristenDemocratisch Appèl—Christian Democratic Appeal
CU	ChristenUnie—Christian Union
D66	Democraten66—Democrats66
fte	Fulltime employment units
GL	GroenLinks—GreenLeft
MP	Member of Parliament
MEP	Member of European Parliament
n	number of cases
n.a.	not available
ns	not significant
PTA	People Talking About
PvdA	Partij van de Arbeid—Labor Party
PvdD	Partij voor de Dieren—Party for the Animals
PVV	Partij voor de Vrijheid—Party for the Freedom
SGP	Staatkundig Gereformeerde Partij—Reformed Political Party
SP	Socialistische Partij—Socialist Party
US	United States of America
UK	United Kingdom
VVD	Volksprartij voor Vrijheid en Democratie—People's Party for Freedom and Democracy
50Plus	50Plus—50 plus

PART I

Background

CHAPTER 1

Introduction

Social Media: A Political Revolution?

Recently, scholars and pundits alike have argued that social media—online platforms that allow a user to send, share, and consume information[1]—were crucial to the success of the Arab uprisings (e.g., Howard & Parks, 2012; Shirky, 2011), that they played an important role in mobilizing people in Latin America (e.g., Harlow & Harp, 2012), and that President Obama to a large extent owed his wins in the 2008 and 2012 Presidential elections to his team's innovative use of social media (e.g., Agranoff & Tabin, 2011; Bartlett, 2013; Crawford, 2009; Katz, Barris, & Jain, 2013; Pollard, 2013; Swigger, 2013). While these examples are highly diverse, all focus on the supposedly high impact of social media in politics, on the instances where social media made a positive difference.

Yet this revolutionary characteristic of social media is far from uncontested. The "utopian view" is opposed by a strong collective of scholars and pundits sometimes labeled as "social-media skeptics" (see Larsson & Svensson, 2014). These skeptics point out the failed social-media protests in Belarus (2006) and Iran (2009) (e.g., Schectman, 2009; Shirky, 2011); show that social media were far from sufficient to topple authoritarian rulers in the Middle East (e.g., Wolfsfeld, Segev, & Sheafer, 2013), claim that social media are no solution for disengagement (Valenzuela, Park, & Kee, 2009), assert that politicians are far more active users of social media than citizens (e.g., Parmelee & Bichard, 2011; Katz, Barris, & Jain, 2013), and address the fact that politicians in several countries saw their reputations and careers severely damaged when they posted messages "in the heat of the moment" that transgressed social and/or political norms (Jacobs & Spierings, 2015; Lee, 2012).

These examples raise a number of important questions. Can both the utopists and the skeptics be right? Do utopists focus on the exceptional positive cases too much? On the one hand, maybe the actual impact of social media in normal politics is far more modest? On the other hand, one could wonder whether skeptics are perhaps too skeptical. Might social media actually change political dynamics or the balance of power? These are the questions that triggered the writing of this book. We will therefore not focus on the exceptions to daily life—the groundbreaking, extreme, and rare events broadcast on CNN, BBC, ARD, or NOS. We will focus on "normal politics" and the way social media can and cannot—or do and do not—reshape the dynamics and balance of the political system in established Western democracies.

Whereas the political struggles in many non-Western regimes are about the power balance between citizens and the authoritarian or pseudo-democratic regimes per se, in Western democracies' everyday politics, the relationships between different political parties, between politicians within parties, between parties and the media, and between parties and social groups are more prominent. A large part of Western democratic politics revolves around the behavior and ideas of political parties, party officials, and politicians. Their use of, and views about, social media's political importance thus are crucial. Yet so far, the few existing detailed case studies or books on social media tend to emphasize the linkage between voter and politics (Parmelee & Bichard, 2011; Katz, Barris, & Jain, 2013). This book should be seen as complementing those studies, providing additional depth with respect to the supply side of politics: the parties themselves. Obviously, we will not lose sight of the voter-politics linkage, as it is an important aspect of normal politics. However, this linkage is just one element of a broader picture: If social media have a transformative potential to change political power relations and inequalities, whether or not that potential actually materializes, depends on these political actors.

The Benchmark: Social Media in US Politics

Most of the studies on social media address the situation in US politics (Conway, Kenski,& Wang, 2013; Evans, Cordova, & Sipole, 2014; Katz, Barris, & Jain, 2013; Lassen & Brown, 2011; Parmelee & Bichard, 2011; Peterson, 2012) or are limited to Anglo-Saxon, majoritarian political systems in which only one politician is elected per district (Gibson & McAllister, 2011; Gibson & McAllister, 2014; Jackson & Lilleker, 2009; Lilleker & Jackson, 2010). In such majoritarian systems, the gap between large and small parties is extremely wide and (candidates from) smaller parties face an uphill struggle more than anywhere else. A result of this is that these smaller parties have far fewer resources but arguably also more motivation/incentive to use (cheap) social media.[2]

Early campaigning research found that the Internet (Web 1.0) mostly benefited major or large parties, as they had the resources to create good-looking websites with lots of bells and whistles and to buy large databases of email addresses (cf. Margolis & Resnick, 2000). Yet the few studies that focused on how social media were used in everyday politics came to more mixed conclusions. Most did at first find some indications of the power balance shifting in favor of the minor parties and their candidates when Web 2.0 started to supplement websites, even though later on, when social media stated playing a more independent role, a return seemed to take place to business as usual. Lassen and Brown (2011), for instance, showed that minor party candidates used Twitter the most, and two years later Conway, Kenski, and Wang (2013) still stressed that nominees from major parties were not the ones tweeting the most. In 2014, however, Evans, Cordova, and Sipole observed that third-party candidates were far less active on Twitter but used the platform differently in that they are (or remain) more interactive and share more personal information. In short, adding up the American studies suggests that technological innovations go through a diffusion process whereby the initial advantage of smaller parties is washed out in later phases. In these later phases there are still differences between parties even though these are best described as "change within a continuity."

Moving beyond US Politics

But do these findings hold true in other contexts? While US politics is highly relevant—it influences many other countries through the Americanization/professionalization of politics (Gibson & Römmele, 2001; Plasser & Plasser, 2002)—the political and media constellation of the United States is rather unique. For instance, US politics is extremely personalized; the political system is almost purely first-past-the-post, leading to a two-party system, and as a result of liberal campaign-financing rules, the big parties' financial resources are unrivaled (Bowler, Donovan, & Van Heerde, 2005). In other words, the results of studies on the United States may well not apply to Western democracies that have a completely different political system and culture.

The central question of the book therefore is: *How have social media transformed normal politics in Western democracies?* We specifically focus on how social media may transform existing power balances in party politics and on whether they mitigate existing inequalities or rather reinforce them—hence the title of the book. By doing so we embed ourselves in the broader theoretical debate that pits those who believe that technological innovations can level the playing field ("equalization") against those who believe these innovations only empower the powerful ("normalization"). Empirically, we will zoom in on the Netherlands and compare it to the radically different case of the United States. Since a lot of the available material deals with the United States, we still do not know whether these insights travel to other Western democracies. By making the Netherlands our main empirical focal point, we can answer that question.

Our contribution is therefore threefold. First, we update and reiterate the theoretical debate about the transformative power of social media and offer a new theoretical framework, namely a motivation-resource-based diffusion model. Second, we expand the scope of the classic theories and apply this model not only from a perspective merely focusing on the competition between parties, but also from a perspective that considers competition within these parties and between different political arenas such as the local versus the national. Third, we test insights from the US literature in a completely different setting, the Netherlands, and offer rich and detailed information about the use and effects of social media on normal politics, placing these findings in a broader comparative perspective.

Theoretical Challenges: Equalization versus Normalization

Time and again the advent of a significant new technology raises the question whether such innovations constitute a so-called game changer (Chadwick, 2013). Social media has recently attracted a lot of scholarly attention addressing that very question (Gibson & McAllister, 2011; Gibson & McAllister, 2014; Gibson, Römmele, & Williamson, 2014; Koc-Michalska, Gibson, & Vedel, 2014; Koc-Michalska et al., 2014; Kruikemeier, 2014; Small, 2008; Spierings & Jacobs, 2014; Utz, 2009; Vergeer & Hermans, 2013). Given the importance of elections in democracies and given that new media and political campaigns are at the heart of both political communication (Graber & Smith, 2005) and comparative politics (Boix & Stokes, 2007), one question that has become particularly salient in both these fields is whether social media replicate existing power differences and mainly benefit the already dominant political actors (the so-called normalization thesis) or whether technological innovations and new media in fact level the playing field, making it easier for minor and marginalized political actors to gain power (the so-called equalization thesis). This "[e]qualization versus normalization is a key debate in the cyber-campaigning literature" (Small, 2008:52), and the question of whether and how new web technologies influence the power balance between parties has now been puzzling scholars for almost 20 years (Castells, 1996; Cornfield, 2005; Gibson & McAllister, 2014; Gibson & Römmele, 2001; Jackson and Lilleker, 2011; Margolis, Resnick, & Wolfe, 1999; Negroponte, 1995; Schweitzer, 2011; Sudulich & Wall, 2010; Vaccari, 2008).

Proponents of the *equalization thesis* suggest that new technologies such as social media level the playing field and redistribute the power balance in favor of previously disadvantaged parties (e.g., Gibson & McAllister, 2011). Regarding social media, the core argument is that social media are (1) cheaper, (2) require less expertise, and (3) allow disadvantaged parties and candidates to bypass traditional media (e.g., Gibson & McAllister, 2014; Vergeer & Hermans, 2013). Furthermore, it has been claimed that its *interactive* nature and its possibility of *anonymity* also benefit marginalized groups (Vergeer & Hermans, 2013). Others espouse the *normalization thesis* and contend that new technologies merely reinforce existing inequalities (e.g., Jackson and Lilleker, 2011). Normalizationists, generally refer to three explanations: (1) online technologies simply replicated old power inequalities because larger parties have *strategic departments*; (2) already powerful politicians are generally better campaigners and more *professional*, and thus better at taking advantage of new technologies; and (3) leading political actors tend to have the resources and *motivation* (see Small, 2008:52).

As will be discussed in more detail in chapter 2, the current normalization-equalization debate, which originated in the early days of the Internet, is in dire need of an update: first, because it is still unclear *why* social media specifically would have an impact on politics; second, because we have little insight in where and when which social media have an impact; and third, because we know very little about which political power relations and inequalities social media have an impact on.

Why Would Social Media Reshape Political Relations?

The theoretical explanation regarding why social media may have an impact needs to be refined. Currently, the impact of *social media* (as something different from traditional websites) is unclear, and there seem to be inconsistencies in the theories. For instance, if social media are cheaper, why then are resources still so important (cf. Gibson & McAllister, 2014; Small, 2008)? And, if social media can be used to surpass existing media, why then do many politicians seem to target their tweets to journalists in particular (Peterson, 2012)? These examples illustrate that it neither has been thought through how the attributes that make social media unique influence politics nor whether politicians' actual social-media usage fits this theoretical uniqueness. In this book we address both these crucial questions.

First, social media (also called Web 2.0) are said to be different from Web 1.0 in several respects. Social media create large networks that diffuse political messages across societal strata through "likes," "shares," and "retweets" at an unprecedented speed and volume— going viral (Steinfield, Ellison, & Lampe, 2008). They also facilitate unmediated and interactive forms of communication that can be used by anyone, given that the platforms are free to access (Boyd & Ellison, 2007; Spierings & Jacobs, 2014; Vergeer, Hermans, & Sams, 2013). The general logic in the equalization and normalization theories is based on the rise of the Web 1.0, but social media have unique characteristics. It specifically needs to be thought through how exactly social media's "causal characteristics" (Goertz, 2006) potentially benefit one group over another. Yet even that is not sufficient: We need to take the debate one step further and incorporate politicians and parties as important *agents* in the framework. It is the actors that enable this causal potential, not just the technical characteristics of social media. Keeping the politicians and parties out of the picture keeps the normalization-equalization debate an academic one, which hardly helps to understand, explain, and inform everyday politics. A few scarce studies focusing on the motivation of politicians is just a first step in this respect (see Small, 2008:52).

Second, the existing literature mainly focuses on how Web 2.0 differs from 1.0. This approach obscures the differences between different social-media platforms. Juxtaposing the two politically most prominent platforms, Facebook and Twitter, shows just how flawed this is. Indeed, Facebook uses far more complex algorithms to punish "bad messages" and is less publically accessible, yet it allows for more visual and personal material. In short, it is a fairly complex personal peg board. Twitter, however, is openly available and a treasure trove for journalists; it is simple but overwhelming, due to it featuring lots of short shout-outs. Twitter is, in sum, a personal press agency.[3] In other words, these different platforms have different strengths and weaknesses, and are perceived and used differently. Understanding *the* impact of social media on everyday politics implies that we need to think through and study how different social media differ, not focus only on Twitter (e.g., Jacobs & Spierings, 2014; Kruikemeier, 2014; Peterson, 2012; Vergeer & Hermans, 2013; Vergeer, Hermans, & Sams, 2013) or lump the different platforms together (e.g., Hansen & Kosiara-Pedersen, 2014; Koc-Michalska et al., 2014).

Where and When Would Social Media Reshape Political Relations?

The second challenge in understanding the impact of social media is knowing where to look. It is quite unlikely that the results obtained from studies on the United States and other majoritarian political systems translate to other nonmajoritarian countries one-to-one. For instance, some claim that politicians can be expected to benefit more from social media in majoritarian systems because the nature of both social media and the system is more geared toward the person, not the party (Vergeer, Hermans, & Sams, 2013), but this has not been studied. At the same time, in countries with relatively strict campaign-financing legislation, social media might have more impact as they are a very cheap campaign tool. To some extent these are empirical issues, but evidently they also raise theoretical questions about context dependency and the generalizability of country-study findings. We will now discuss three such questions.

First, the political system can be expected to shape the extent to which social media affect the political power balance. Considering the characteristics of social media, the degree of personalization, the campaign-financing regulations, the media system, and the role of journalists are our main focus here. The mechanisms behind social media's impact on the power balance in politics between the different actors should be assessed against that background. However, as said, a lot of the established Western democracies are parliamentary systems with a proportional electoral system in which the dominant position of the biggest parties is more precarious to begin with. Such a context may increase the zeal and opportunities of the smaller parties in applying technological innovations, while bigger parties at the same time have to be on the lookout, as their position is more vulnerable.

Second, social-media theories are usually tested at the national level. As such "first-order" elections attract a great deal of attention from citizens and media alike, one can expect that the added value of social media is relatively limited. It is likely that in second-order elections, such as local or European elections, the unique potential of social media to create large networks has a stronger transformative power, given that these arenas often attract far less attention from the traditional media and thereby often largely remain invisible. Unfortunately, these two arenas are often neglected (Gibson & McAllister, 2014:2; though see: Larsson, 2013; Vergeer, Hermans, & Sams, 2013). In other words, to understand the impact of social media in nonnational elections, we argue for more theoretical attention to the way the transformative power of social media is conditioned by the visibility of a political arena.

Third, in many countries, including the Netherlands and the United States, social media are no longer a complete novelty, and their impact might have changed over time. If pioneers have shown that social media improved their situation vis-à-vis the frontrunners (equalization), the latter, having seen what works and what does not, might have decided to pick up the glove and start using social media too (see Gibson & McAllister, 2014). Moreover, in the early-adoption phase, politicians who used social media might have benefited from being modern and progressive.[4] Yet, as soon as social media became part of everyday politics this advantage is likely to have evaporated. In fact, the effect may be such that those who shy away from using social media are at a disadvantage: the early-bird benefit has then changed into an absentee penalty. In sum, the timing and political popularity of social media needs to be assessed to understand its actual equalizing potential.

On Which Political Power Relations and Inequalities Do Social Media Have an Impact?

The final theoretical (and empirical) challenge can be found in the lack of an overarching framework. The equalization-normalization framework might be argued to be the most comprehensive one, but it mainly focuses on interparty relationships; most research on the transformative power of social media primarily examines how social media change the dynamics between (candidates of different) parties, studying whether major, minor, or fringe parties benefit more from technological innovations in communication (though see: Karlsen, 2011; Vergeer, Hermans, & Sams, 2013). As many would argue that the power balance between parties with different ideologies is most indicative of the direction a country takes, this focus might appear logical, but there is actually no theoretical reason not to expect social media to shape intraparty relationships—the power balance between different politicians in a party—or those between the individual politician and the party.

Given the logic just mentioned, intraparty relations might seem trivial, but nothing could be further from the truth. These power balances are crucial in shaping the degrees and practices of personalization and representation, and thus fundamental to the functioning of our democratic systems. If social media allow individuals to build large personalized and unmediated networks, this might actually be most useful to politically marginalized and underrepresented groups such as women and ethnic minorities. We do know that the first woman on a list and the first ethnic-minority candidate often receive a bonus in list-proportional electoral systems (André, Wauters, & Pilet, 2012; Thijssen & Jacobs, 2004), and in the United States the "social-media president" (Katz, Barris, & Jain, 2013) was also the first African American President. Our knowledge about how ethnic or gender profiling changes the intraparty power balance, however, is fairly limited.[5] Politicians representing specific groups such as ethnic minorities or women can make use of the nonmainstream profiling and tailor to these specific groups without being bound by geographic distances (Blumler & Kavanagh, 1999; Enyedi, 2008).

Similarly, there might be diffuse personalization—strengthening the power of individual politicians vis-à-vis parties (Van Aelst, Schaefer, & Stanyer, 2011)—because politicians now have cheap access to their own campaign tools as well as to direct communication channels with journalists on social media (Lasorsa, Lewis, & Holton, 2012; Vergeer, Hermans, & Sams, 2013). As such, social media can also undermine the centralization of the party and make individual candidates more independent from the party leadership.

In other words, the transformative potential of social media can well be expected to extend beyond the interparty relationships that are so central to the equalization-normalization debate. Not adding that important new dimension could easily lead to underestimating the impact of social media. Moreover, given that social media are typically personalized communication tools (Vergeer, Hermans, & Sams, 2013), we are actually more likely to see their impact on intraparty politics than on interparty politics.

Research Design

In the previous section, we formulated three theoretical challenges in developing further understanding of the way social media can reshape and are reshaping everyday Western politics, particularly power relations or political inequalities. The second chapter of this book will address these challenges and formulate our motivation-resource-based diffusion model. In the chapters after that, we will test the expectations derived from this model.

In this next section, we will discuss how our case study from a comparative perspective is set up and which data sources we will use to answer the core questions formulated above.

A Case Study From a Comparative Perspective

As mentioned earlier, more research is needed to establish whether the results found for the United States translate to other Western democracies, and, more generally, to find out how the political context influences the impact of social media. This book faces that empirical challenge by studying the Netherlands from a comparative perspective. As will be clear by now, we define a "case study in comparative perspective" in line with Gerring (2007:19), as "an intensive study of a single case [a spatially delimited phenomenon observed over a period of time] to shed light on a larger class of cases." Specifically, we will study the Netherlands from about 2009–2010 to 2014–2015, explicitly providing empirical information on the larger class of cases. This case study is thus an in-depth examination of the how and why: How and why do social media reshape the power balance between parties and politicians? The focus therefore is on within-case variation. This within-case variation in our study can encompass variation in the extent to which politicians use social media and the variation in the parties' campaign strategies, the popularity of social media over time, the types of messages posted, the reasons for using social media, the risks and opportunities associated with social media, the journalists' treatment of social media, between political arenas, the resources politicians and parties have, the ideologies, and so forth and so on.

One of the major advantages of this approach is that we can combine a theory-testing and a theory-generating approach (Gerring, 2007:39–41; see also Beach & Pedersen, 2013). On the one hand, we can test the mechanisms proposed by equalization and normalization theory not only in terms of whether the expected effects (changes in the power balance) are found but also whether these changes occurred for the reasons those theories suggest. On the other hand, given that we apply the theories to intraparty relations, in different contexts, and at new levels, it would be very limiting to only test those two theories. In other words, we should also be open to new ideas and political mechanisms arising from the study, as some elements have hardly been theorized before.

The case-study approach allows us to combine theory testing with theory generating, and this is highly suitable for a field such as this, where the existing theory could benefit strongly from further development. As such, our focus is on internal validity. Our choice not to focus on one point in time but to instead cover a somewhat longer time period also helps us in this respect. The time period covered in this book includes two national elections (2010, 2012), as well as local and European elections (2014). Moreover, though social media were a political novelty in 2010, they quickly became widespread in the years that followed. This focus increases our within-case variance and helps us to track developments that might influence the impact of social media on political power relations.

Last but not least, since we place our case study in a larger literature and explicitly relate it to existing studies on other countries, such as more similar European political systems or the United States, we can reach a larger external validity than a classic, independent case study.

The Netherlands

Among the established Western democracies, the Netherlands is in many ways a counterpoint to the United States, and not just because there are roughly 20 Americans for every Dutch citizen and 220 square miles of US territory for every square mile of the Netherlands. The Dutch political system is anything but majoritarian: highly proportional, it has roughly a dozen parties in parliament, and the most proportional translation of the percentage of votes to the number of a party's seats in parliament (Farrell, 2011). In addition, campaign budgets have been fairly limited and, when compared to the United States, the central parties are far more powerful than individual politicians (Andeweg & Irwin, 2005). Hence, if results found for the United States also hold in the context of the Netherlands, this would strongly suggest they are likely to be found elsewhere as well (echoing the case-study logic of a most different system design). Or to put it in fewer than 140 characters: If we can find the same results there, we can find them anywhere.

This is not the only reason why the Netherlands is an interesting case. The country has also been a digital frontrunner in Europe; it is roughly two to three years ahead of most other European countries. The Dutch have a top-ten Internet-penetration rate worldwide—94 percent since December 31, 2013 and rising[6]—as well as a top-ten broadband Internet subscriptions—39.4 percent in the early 2010s.[7] The social-media penetration is high too, with 3.3 million Twitter users in 2012, in a population of approximately 17 million (Oosterveer, 2013). According to some sources, the Dutch are also the most active social-media users worldwide (Dugan, 2012; Woollaston, 2013). Politicians are also very present on social media: 76 percent of the 531 candidates of the 11 biggest parties in the 2012 national elections had a Twitter account, and 72 percent were present on Facebook.

This frontrunner position makes the Netherlands a particularly interesting case for at least two reasons. The first reason is societal and political, studying this case in detail can help make predictions and formulate policy advice for countries that will probably go through similar developments. The second is that a country with a longer history of social media use can help show developments over time. As suggested by the literature, the initial advantage smaller parties have may later wane, and to study this, we need to be able to look at developments over time.

More generally, zooming in on one country from a comparative perspective allows for more in-depth analyses without losing sight of the larger picture. This particular comparative perspective not only includes a contrast of two extreme cases (the United States vs. the Netherlands), but the Netherlands is also a good archetypical case within the European family of consensus democracies (Jacobs, 2011), and its multiparty system and ballot structure are fairly typical for most Western democracies (Farrell, 2011; Colomer, 2011).

Data Sources

This book primarily draws on expert interviews, aggregate election data, and social-media data from the VIRAL project.[8] Over the course of the project, we conducted 18 interviews with high-profile politicians and with the social-media and web managers of nine parties represented in parliament. These expert interviews provide unique and valuable "behind-closed-doors" information about how political actors perceive the uses, dangers, and opportunities of social media, as well as factual information on the available resources and party policies and control. This qualitative information is accompanied by hand- and computer-coded quantitative data on all 1,024 candidates of parties that were represented in parliament with at least one seat. Of these candidates, we know whether they were present on Twitter (2010/2012), the Dutch social-media platform Hyves (2010), and Facebook (2012). For some years and platforms we also have information on how many friends or followers they had, and how many messages they posted in the period leading up to the elections. For the 150 Members of Parliament (MPs) in 2010, 2012, and 2015, we have the same or even richer data available. In addition, demographic (gender, ethnicity, age) and political (media coverage, incumbency, preference votes, party, list position) data are available. The core focus of these data is on the supply side of politics: which politicians and parties have decided to use social media and to what extent they did. For several nonnational elections similar data is available.

The analyses of these data are accompanied by additional information derived from social-media content, media coverage, election surveys, party documents, and secondary sources (existing studies on the Netherlands). These data help to unpack the link between politics, social media, and "old" media; to disclose politicians' views (as expressed in the media); to sketch the electoral background to our core questions; and to illustrate politicians' presentation and behavior on social media. Evidently, we cannot definitely answer all our questions, certainly not as our study is partly about generating theory, plus there are limitations to the data available. In this respect, our decision to focus on the relationship between parties, politicians, and the media more than on the linkage between politics and voters also requires more focus on the sources used in this book than on voter surveys, for instance (cf. Katz, Barris, & Jain, 2013; Parmelee & Bichard, 2011; Schlozman, Verba, & Brady, 2013).

Roughly speaking, we can compare a case study to a murder case or jigsaw puzzle—as has been done in the literature (e.g., Van Evera, 1997)—in the sense that we have to piece together the evidence and weigh bits and pieces of information, which can be anything from in-between a doubly decisive and a straw in the wind piece of evidence (see Beach & Pedersen, 2013; Gerring, 2007; Van Evera, 1997). These pieces together enable us to piece together the larger picture but will also help to pinpoint the pieces that are still missing. To some extent we already know what these will be: broad interpretative or qualitative analyses of the social-media posts of many politicians and parties are rare, for instance, partly because this is a very laborious endeavor, and such systematic content analysis is beyond the scope of this book as well. Moreover, we first need to try and solve our puzzle or murder case before we can more definitely tell which pieces are missing. For instance, a broad systematic qualitative content analysis of social-media profiles and posts would benefit from a spotlight on the *type* of content that is crucial for finding the missing links.

Overall we draw from various data sources, which is natural for a case-study approach, and this book brings together a unique collection of data on the Netherlands. Where relevant, more information on the exact data and their collection process is provided in the text.

Techniques

The techniques used to analyze the data depend a great deal on the specific data, and we will therefore discuss these techniques in greater detail in the chapters where they are applied. They encompass everything from qualitative content analyses and the close reading of interview transcripts, to statistical group comparisons and multilevel or (negative) binomial regression analyses.

Outline of the Book

The questions, theories, data, and analyses discussed above will be presented in the seven chapters that make up the three parts of this book. Part I consists of three chapters (including this introduction), providing the necessary background information. Part II presents three chapters that each provide empirical analyses of the Dutch case from a comparative perspective and concern themselves with a specific element of social-media equalization and normalization in normal politics. Part III finalizes the book and reflects on the overall results in two chapters, asking whether social media have a tangible transformative impact and what the road ahead looks like. Below we give a more detailed outline per chapter.

Chapter 2 provides the theoretical background, presenting a systematic overview of the unique characteristics of social media as a means of communication and as a campaigning tool, and relating these to the existing theoretical debates about equalization and normalization. It then explores how and why the transformative impact of these unique characteristics may differ by political system, arena, and political diffusion of social-media context. We end the chapter by extending the debate from the power relations *between* parties to the relationships *within* parties. In closing, this chapter introduces our new theoretical framework, the motivation-resource-based diffusion model.

Chapter 3 outlines the empirical background, discussing the political system of the Netherlands as a typical multiparty parliamentary system and comparing it to other European and North American democracies. It offers general information about the role of social media in politics and about the data on social-media usage by political candidates, as well as newspaper coverage of social-media usage from a comparative perspective. Above all, this illustrates the increasing importance of social media as perceived by politicians and political journalists, but it also reveals some discrepancies between the opportunities social media offer and their actual usage in politics. At the end of the chapter, we complete the picture by making the connection to the voters. Looking at the latter suggests that the broader public does not share the enthusiasm of the politicians and journalists. Citizens are slower in adopting social media, indicating that social media is most likely to have an indirect political impact (i.e., via traditional media) in the current situation.

The first of the three chapters in Part II, chapter 4, focuses on inter-party relations: Who benefits most, the major or minor parties? Using our interviews and social-media data for all Dutch parties in parliament, we show that it is crucial to take the parties' ideological signature (e.g., postmaterialist, populist) into account. Doing so shows that grouping minor parties by ideological characteristics better explains their social media usage and that minor parties are not a homogenous group for which social media has one and the same effect. Moreover, there is a considerable difference between the situation early on, in 2010, and the time when social media had proven themselves in the eyes of the major parties. Overall, we specify how in the end the unique combinations of the different resources available and dedicated to social media and the (de)motivation behind its use make social media beneficial to the major parties and the more postmaterialist minor ones, while the populist and traditional nonmajor parties are mostly disadvantaged. Chapter 5 deals with the intraparty relationship, arguing that social media can be expected to weaken the position of a party's main politicians and its central leadership, and (thus) particularly benefit politically underrepresented groups. Departing from this point, we first look at the way parties deal with the personalized campaigning that social media induce and the quantitative and qualitative diffusion of social-media usage within parties among different ranks of politicians (list pullers, top 10, and lower ranked politicians). We also specifically study the behavior of women and ethnic-minority candidates, their presence on social media, and their usage and identity presentation. Empirically, social media do not live up to their personalization and profiling potential here: dominant politicians remain dominant, and there are only few differences in terms of gender and ethnicity. Only few high-profile candidates manage to use social media at a high-quality level, and these are mainly the more postmaterialistist candidates who have a network and expertise already. Still, both the party management and politicians seem to realize that the larger potential is there. In chapter 6, we look at the role of the issues dealt with in chapters 4 and 5 in sub- and supra-national politics, providing new material on both local and European elections. The results presented in this chapter roughly show the same results as the previous chapters, but they also refine them and help us understand the dynamics in these political arenas in great detail.

Part III starts with chapter 7, posing a final question that reflects on the results from Part II: Do social media help win elections? The Obama campaigns are examples of the tremendous potential of social media, but the generalizability of these two elections is not yet clear. We argue that social media may not make people shift their vote from a highly conservative party to a green party, but they can in several ways influence people's voting decision within a certain bandwidth. This argument is illustrated using the preference votes for candidates in a list-proportional system. Our analyses as well as other studies more and more show a clear but small effect, particularly after the early-bird benefit disappeared. Finally, in chapter 8, we bring together the results from the empirical chapters, assess the overall transformative power of social media, and explicitly link this discussion to the debates introduced in chapters 1 and 2. In short, we find a multistage process whereby particularly post-materialist parties and candidates in those—less centralized—parties benefit from social media from the start, because of their enthusiasm and the expertise that is almost intrinsic to their ideology. This is equalization. Later on major parties and politicians catch up and have enough support to do so. They normalize their position vis-à-vis the other than postmaterialist actors. Professional usage becomes more and more important in this stage. When finally the most traditional, centralized, and populist parties start adoption social media more and more, their backlog in experience and expertise is considerable, and they have difficulty catching up. They mainly lose against the other nonmajor parties. Finally, the most innovative actors keep ahead by moving to new platforms or higher-quality use, but the old media logic remains rather dominant and social media seem to be underutilized for community-building and benefitting underrepresented groups.

CHAPTER 2

Theorizing Social Media, Parties, and Political Inequalities

The Debate

The role of social media—often called Web 2.0—has been addressed in a growing number of studies, studies that often build on the literature regarding politics and Web 1.0 (Gibson & McAllister, 2014; Schweitzer, 2011; Vergeer, Hermans, & Sams, 2013). Indeed, the central debate in both the earlier and the most recent literature is whether or not new web technologies influence the power balance between parties. By now this question has been puzzling scholars for almost 20 years (Castells, 1996; Cornfield, 2005; Gibson & McAllister, 2014; Gibson & Römmele, 2001; Jackson & Lilleker, 2011; Margolis, Resnick, & Wolfe, 1999; Negroponte, 1995; Schweitzer, 2011; Sudulich & Wall, 2010; Vaccari, 2008). Some suggest that new technologies, such as social media, level the playing field and redistribute the power in favor of previously disadvantaged parties—equalization (e.g., Gibson & McAllister, 2011). Others espouse that new technologies merely reinforce existing inequalities—normalization (e.g., Jackson & Lilleker, 2011). Most recently, Gibson and McAllister (2014) suggested that equalization occurs first (Stage I), then normalization (Stage II).

Before we can understand the impact of social media, however, it is crucial to have a proper understanding of what social media actually are. In this chapter we will not simply repeat the arguments made by earlier studies but instead evaluate these studies' arguments from the angle of the characteristics that make social media unique and that determine the impact they may have—their so-called causal attributes (cf. Goertz, 2006:5). As Web 2.0 is said to differ from Web 1.0 in several respects (Gibson & Ward, 2012:63), it is important to reflect on these differences and how they influence the potential effects social media might have. In this chapter, we shall start by doing exactly that in the first section before outlining and critically reflecting on the equalization-normalization debate in the next section. Based on this diagnosis and the analysis of social media's unique characteristics, the subsequent section then presents our motivation-resource-based diffusion model. This model lays bare core questions and is the foundation for our expectations about how, why, and where social media influence power relations in everyday politics. We will focus on three relatively new topics in particular: the difference between social-media platforms, the moderating effect of the political context, and social media's impact on intraparty power relations.

Social Media in Politics

Traditional Web 1.0 tools, such as personal webpages, email, and text messaging, are quite different from tools from the social-media or Web-2.0 era (cf. Bimber, 2012:119). Indeed, social media are not just a next step in online communication; they take online communication in another direction, a more personal and interactive one. These unique characteristics should be at the core of our theoretical reasoning about their *potential* effects.

Social Media as Unique Communication Tools

We distinguish five characteristics, the first two of which are also characteristic of Web 1.0 online tools. Social media are (1) unmediated, (2) personal, (3) interactive, (4) cheap and easy to use, and (5) able to go viral.

1. *Unmediated*—Just like Web 1.0, social media provide a direct, unmediated connection with a larger audience. In the case of lesser-known politicians this audience may be limited to their own network of family and friends, but for more well-known politicians, this network can extend quite significantly. Messages posted on social media appear online unmediated. There are no gatekeepers that select which messages are deemed newsworthy, and journalists cannot restructure the messages to frame them differently (Bimber, 2012:120; Polat, 2005; Tedesco, 2008).

While most, if not all, social media have some guidelines on what or what cannot be posted, users are largely autonomous in deciding what they post.

2. *Personal*—Even more than in Web 1.0, social media push the personalization of content and communication (Vergeer, Hermans, & Sams, 2013). Social media are typically linked to an individual and allow for the visualization of a person's private and professional life. As such, social media are an easy way to offer a look behind the scenes. Social media being unmediated largely facilitates this: The information's owner is in control. As a result, organizations and people also have more freedom in shaping their style of communication; they can give their messages a more personal touch (Kruikemeier, 2014). Because of these differential styles, as well as the receivers' sense of being directly connected to the person or organization, the receiver might also experience a more personal connection.

3. *Interactive*—Unlike Web 1.0 tools, the option to involve others in your communication is typically a core feature of social-media platforms (boyd & Ellison, 2007; Spierings & Jacobs, 2014; Vergeer, Hermans, & Sams, 2013).[1] This can be done either through comments on the posts of other users or by pulling a person into a digital conversation using, for instance, Twitter's "@-mention" functionality or "tagging" a person on Facebook. This makes social media fundamentally different from offline media as well as some other forms of online media (see Kruikemeier, 2014; Sundar, Kalyanaraman, & Brown, 2003). Social media makes it possible for people and organizations to become involved easily. They are not merely receivers of the information but also creators, as with one click of a mouse or tap of a finger it is possible to respond, give one's opinion, ask for more information, or draw attention to an interesting piece of information.

4. *Cheap and easy to use*—Building a personal website and collecting a list of phone numbers or email addresses was once quite a cumbersome operation. Indeed, as Gibson and McAllister (2014) note, much of the potential of traditional online tools was diminished by the high level of expertise needed to use them properly. Yet opening a social media account is within reach of almost everyone in Western democracies, not just because of the political freedoms in such countries but also because Internet penetration is very high, mobile devices with broadband access are almost

universally available, and, above all, because opening an account is free (boyd & Ellison, 2007; Vergeer, Hermans, & Sams, 2013). The startup costs are nil, as the revenue model of social media is based on advertizing and users' data.[2] Still, the startup and everyday use of social media are free and accessible.

5. *Virality*—There is a reason why social media are sometimes called social-networking sites (e.g., boyd & Ellison, 2007; Utz, 2009; Vergeer, Hermans, & Sams, 2013): social media connect profiles by befriending or following each other, and the information posted by one user is shared through such a network. Typically people can also see what a person they are connected to "likes" and some posts thus draw attention not only through first-degree followers, but potentially also to second-degree and even more distant connections (Kruikemeier, 2014; Stromer-Galley, 2004). The speed and volume of social media, and their ability to snowball from first-degree friends to friends of these friends is potentially a very powerful and new characteristic of social media. Considering the "six degrees of separation" theory—suggesting everyone can be connected to anyone else on earth in six steps—this means that posts can be diffused to an incredibly large number of people across all social strata with incredible speed (Bond et al., 2012).

Social Media as a Political Communication Tool

Being a revolutionary communication tool does not automatically make social media a revolutionary *political* communication tool. Neither does each of the characteristics above by itself hold extraordinary transformative potential. To understand how social media may reshape politics, we have to theorize how its characteristics interact with each other and how these combinations can change parties' and politicians' communication. Reflecting on different combinations of social media's characteristics suggests at least four ways in which these platforms have transformative potential.

First, having access to an *unmediated*-communication tool that is primarily *personal* and is *cheap and easy to use* makes social media well suited to individual politicians for spreading their individual messages. Moreover, such messages might even go *viral* and directly reach millions of people, party members and affiliates, as well as the general public. This means social media offer excellent opportunities for politicians (and to a lesser extent parties, as social media are personalized) to advertise themselves. It is a simple addition to their communication and campaigning toolbox that not only helps to communicate directly with people but also showcases the politician or party as a modern actor that knows (and wants to know) what is going on in society and that evolves with changes in society. Surely, other mass media also made it possible to reach a larger audience—hence the name *mass* media. But such media worked through gatekeepers and typically added a layer of framing to the original message; traditional press releases seldom appeared in their original form (Harcup & O'Neill, 2001). Overall we label this potential the ***advertisement opportunity***.

Second, while the advertisement opportunity makes use of four characteristics of social media, it leaves out the fifth one—interactivity. The *interactivity* of social media is nevertheless a powerful characteristic. Combined with the unmediated and personal nature of social media and the fact that social media are not only cheap and easy for politicians but also for citizens, interactivity allows politicians to establish contact with individuals. This can be seen as a proxy for human contact, suggesting that a politician is approachable. Even those who do not engage with the politician themselves can see that he or she answers questions or reacts to messages. In short, such behavior suggests a "social presence" (Kruikemeier et al., 2013; Short, Williams, & Christie, 1976). Social media indeed make politicians more approachable: The psychological threshold to sending a tweet or Facebook message may be considerably lower than to write a letter (which is also more time consuming) or call a politician (which may be rather scary). As such, the distance between the two spheres—those of "normal people" and the previously remote politicians'—can be decreased (Kruikemeier, 2014; Spierings & Jacobs, 2014; Sundar, Kalyanaraman, & Brown, 2003). In addition, social media allow politicians and parties to provide more of a peek behind the scenes, into their daily worries and the smaller things on their agenda (Jacobs & Spierings, 2015; Vergeer, Hermans, & Sams, 2013). Doing all this can be very costly when using only television or newspapers ads, so campaigners have to be very selective in what to focus on. In this respect, social media are real addition to the political communication toolbox, as they are especially well tailored for these small bits and pieces of information. To sum up, the aforementioned characteristics of social media allow parties and politicians to connect with people at a different, human level. They can create stronger ties, because voters can more easily relate to them and interact with them. Let us call this the ***human-contact opportunity***.

Third, political communication is often thought about in terms of geographical constituencies, in particular since television and radio are largely organized geographically. More than any other media, social media transcend the geographical limitations of communication. The *unmediated, interactive,* and *cheap and easy-to-use* nature of such media allows parties and politicians to carve out not necessarily geographically defined niches and reach out to very specific social groups that are spread out over the whole country.[3] Examples of such niche groups are people in a certain industry (e.g., social workers, primary-school teachers, or freelancers) or certain social groups (e.g., Muslims, LGBT people, women, or students). Whereas regular mass media might focus on the dominant groups in a society and pay little attention to certain specific groups, these groups can actually tip the balance in an election or can be ideologically important to a party or politician. Moreover, as people also tend to connect with "kindred souls" on social media, such media foster bonding social capital (boyd & Ellison, 2007; Steinfield, Ellison, & Lampe, 2008; Utz, 2009). Politicians and parties can use social media to get inspiration from their followers or strengthen and expand their community. We call this the ***target-group opportunity***.

Fourth, the *personal* nature of social media communication, their *interactive, low cost, and user-friendly* nature not only mean that traditional media are no longer (the only) gatekeepers, they also mean that journalists themselves can be targeted more easily by individual politicians and parties. Conversely, social media are an easy tool for these journalists to monitor politics. Though some journalists claim that finding valid political information might be "like searching for medical advice in an online world of quacks and cures" (Goodman, 2009), most of them are regularly looking for quotes and one-liners on policy positions rather than factual information (Peterson, 2012). Social media caters to this desire for "quotable" information and as such may offer politicians and parties a "new door into an old house," a means of getting into the traditional mass media. Social media, and the Internet more in general, are somewhat analogous to the eighteenth-century salons this way, as aptly observed by Herbst (2011:95). Listening to what was said in these salons meant listening to "conversation [that] reflected and shaped the culture of France and much of Western Europe and ignited the revolutions that would change our world forever." In salons, the elite came together and discussed societal issues. Social media connect politicians and journalists, hence we use the term ***salon-debate opportunity*** (Table 2.1).

Table 2.1   Overview of the four opportunities social media present[a]

1	Advertisement opportunity	Sending out unfiltered messages with the potential of reaching a wide audience
2	Human-contact opportunity	Creating a feeling of social presence by reacting to and interacting with followers
3	Target-group opportunity	Reaching out to geographically dispersed niche groups
4	Salon-debate opportunity	Contacting journalists directly and debating with them

[a] Theoretically no fewer than 32 combinations of the five characteristics can be constructed, but not all of these combinations are relevant for social media. Based on the expert interviews and a review of the available literature, we delineated the current four opportunities. Without a doubt others can be considered too, but these are the most salient and most likely to have an impact. If these four opportunities do not have an impact, one can safely say that social media are unlikely to change everyday politics.

The Dark Side of Social Media: Social-Media Malaise?

So far we have only spoken about opportunities, but obviously social media can also *harm* politicians, parties, or even democracy itself. Some social-media skeptics such as Katz, Barris, and Jain (2013), Parmelee and Bichard (2013), and Shirky (2011) voice concerns about the impact of social media on politics.[4] The five characteristics of social media may well have negative consequences too, some of which may be minor, while others could be more substantial (or less likely).

The *unmediated* nature of social media certainly entails the risk of politicians sending out messages in the heat of the moment, messages that then can never be removed from the Internet. This may damage the reputation of parties and politicians alike. Yet that risk was always present; slips of the tongue make for great news as they often more readily meet the criteria of the newsroom (Harcup & O'Neill, 2001). The only additional downside to this phenomenon on social media is that, as such messages can now go *viral,* their damage may also be done more quickly. However, when the speed increases, simply surviving a "Twitter storm," for instance, may be enough for a politician, as the next day another message may have gone viral, replacing the politician's slip of the tongue as the center of the attention.[5]

The *cheap and easy-to-use* nature of social media allows politicians to try out social media, creating their accounts, experimenting a bit, and then abandoning them. Such "social-media zombies" may make for quite hilarious moments and can indeed damage the reputation of individual politicians (see Wilson, 2009),[6] and merely having a social-media account will surely not be enough reap the (supposed) benefits of social media (Crawford, 2009), but it is not certain whether that would actually damage these politicians. Probably not, as they often have too few followers for anyone to notice. Surely, such politicians may occasionally find themselves the center of attention, but such instances are more likely to be exception than rule.

What about the *interactivity* of social media? Research on British and Dutch MPs' early use of Web 2.0 suggested that politicians did not change their habits and were hardly interactive on social media (Graham, Jackson, & Broersma, 2014). By now this may have changed. Some politicians may even have become too interactive, trying to answer every comment (cf. infra, Pvda, MP, 2013). Surely, this may make them seem embittered or see them accused of being a know-it-all, but it remains to be seen how many politicians act this way.

Lastly, the fact that social media are *personal* communication tools may speed up or facilitate the personalization of politics, which in turn may trigger an increased focus on the private lives of politicians rather than on their policy and ideological positions. Such fears echo those voiced by scholars adhering to the "media malaise" thesis (for an overview, see Norris, 2000). If social media become the dominant communication tools, this may result in fact-free politics, which in turn may increase cynicism about politics and even dissatisfaction with the way democracy works. At its worst, social media may thus trigger a "social-media malaise." Whether or not such a malaise is likely hinges on many prerequisites: to begin with, politicians and parties must use social media in a personalized way; additionally those messages must focus on their private lives rather than on matters of substance; in addition to which citizens must consume such messages instead of other sources of information; on top of which such consumption must also trigger negative feelings and politics. All in all that is quite a long causal chain. In fact, the evidence on the original media-malaise thesis is mixed, suggesting both negative and positive effects (Curran et al., 2014). While it is too early to examine the social-media malaise thesis due to lack of data availability, for now it seems unlikely that social media would have such pervasive consequences, given that in many countries traditional media are still the dominant source of information.[7] All in all it thus seems that social media offer politicians more opportunities than pitfalls.

Moving beyond Mere Opportunities

Having discussed the opportunities by which social media can reshape everyday politics, a few remarks are in order. First of all, these opportunities mainly concern the communication between politicians and citizens, not so much on the results of this communication. Making a (sustainable) connection with citizens is often the first step in a causal chain. After making a connection, politicians can start considering whether or not to ask people to vote for them, or for instance use social media as a crowd-funding tool, like Obama did (Spierings & Jacobs, 2014). Furthermore, the opportunities themselves do not always materialize. Barack Obama can be cited as the first to have genuinely mastered the potential of social media, and the Obama campaign was especially successful in mobilizing his grassroots supporters (Zhang et al., 2010), but by no means does this mean that realizing this potential is a given. Social media, for instance, make it easier than ever to contact journalists directly, but it may well be that politicians and parties do not make use of this opportunity, or that journalists simply ignore those messages. The opportunities thus need to be integrated in a broader theoretical framework about the uptake and use of social media in order to assess to what extent equalization and normalization can be expected—who benefits from social media and when? In the above sections, we have tracked the possibilities offered by social media. So far these have been discussed generally, as if they exist for all parties and politicians. However, for some politicians and parties the opportunities might better fit the already existing culture, background, electorate, and so on than for others. Ultimately, whether the opportunities discussed here fulfill their potential to reshape the power balance between parties is of course a matter of empirical assessment, something we will embark on in the empirical chapters of the book.

All in all, this section leaves us with three questions: (1) do politicians and parties perceive social media as communication tools offering the unique opportunities described above? (2) Are social media used in line with their unique potential? And (3) how can these opportunities reshape the power balance between major and minor political actors? The first two (empirical) questions will be addressed in chapters 3 to 6. The last question is the central theoretical question we will now turn to.

Revisiting the Equalization-Normalization Debate

How can social media's communication opportunities reshape the power balance between major and minor political actors? To answer this question, we will go back to the clash between equalization theory and normalization theory and first present the state of affairs in that debate. It will turn out that both theories' theoretical foundations are rather underdeveloped. Based on our more detailed discussion of social media's characteristics above, we will then flesh out these theoretical positions and formulate our core expectations.

Equalization or Normalization?

Studies adopting or testing the *equalization* perspective generally use the following arguments. As smaller parties have fewer resources and need to bypass traditional media (e.g., Vergeer & Hermans, 2013; Gibson & McAllister, 2014), they benefit from the unmediated and cheap and easy-to-use nature of social media. It has also been claimed that the interactive features of the technology benefit postmaterialist parties, and, as these tend to be minor parties, this is an argument in favor of the equalization thesis.

How convincing are these three explanations? *If* social media are cheaper and require less expertise to have the same (or greater) reach and penetration than traditional media, then surely smaller parties, with smaller budgets and therefore fewer possibilities to hire consultants or in-house experts, should benefit the most. The other two equalization mechanisms are less straightforward. Smaller parties are often less visible in traditional media, so bypassing traditional media may give them a greater reach, but there are two important caveats: larger parties can also use social media to bypass traditional media, leading to normalization if they then dominate social media as well, as journalists can simply ignore the smaller parties' social-media appeals. Similarly, there is no reason why mainstream parties could not benefit from the interactive features of social media and try to establish a sense of social presence. Surely a party does not necessarily need to be postmaterialist to use interactivity. In fact, the interactive features may actually be most advantageous to larger parties, as they have more personnel capacity to respond to large volumes of comments and questions, and have larger numbers of party activists who can be "excited" and motivated to campaign for the party both off- and online.

Overall it seems then that the equalization argument is based on an implicit assumption that large parties are simply less inclined to use social media; if they do decide to use them, some of the earlier arguments would suggest normalization might be more likely to follow than equalization.

Generally, the explanations provided for a *normalization* effect focus on larger parties having more resources and thereby being better placed to make use of the aforementioned opportunities. Indeed, larger parties are said to have strategic departments and professional politicians (Gibson, Römmele, & Williamson, 2014; Small, 2008; Vergeer & Hermans, 2013). And this rationale echoes classic sociological theories predicting that technological development leads to increasing inequalities because of resource build-up (cf. Lenski, 2005). In addition, Small claims that "[m]ajor parties have the resources and *motivation*" (Small, 2008:52, emphasis added). Where this "motivation" comes from, however, is unfortunately hardly specified.

How convincing are these mechanisms? Having strategic departments and more professional politicians can help any type of campaign. Yet why would resources be such an advantage? Or more generally, the expectation that normalization occurs because bigger parties "were prepared to invest in it" (Gibson & McAllister, 2014:1) or "have the resources and motivation" (Small, 2008:52) is fairly puzzling, as such claims directly oppose the characteristic of social media that they are cheap and easy-to-use. Moreover, the normalization argument that parties are *prepared* to invest in it, Small's "motivation" (2008:52), raises the question why bigger (e.g., catch-all) parties would have more motivation to use social media than small ones? After all, shouldn't one say that smaller parties have more incentives to invest in innovations because they lag behind and need to try to catch up?

The Motivation-Resource-Based Diffusion Model

In a recent study, Gibson and McAllister (2014:4) integrated both perspectives for the Australian case. Based on "cumulated snapshots" and new analyses of Australian elections, they suggest cyber-campaign innovations follow a multistage process. The mechanism the authors provide builds on the ones outlined above: Smaller parties have incentives to experiment with innovative tools, which gives them an initial advantage if the tools turn out to be successful. In such cases "larger parties saw the added value of online campaigning and were prepared to invest in it," and the interparty relations will normalize again (Gibson & McAllister, 2014:1). This framework is an important step forward, as it accounts for inconsistent earlier findings. Unfortunately, the underlying mechanisms still remain largely unknown and the resource-arguments remain at odds with the nature of social media. Theoretically, Gibson and McAllister's framework needs further refinement, specifically regarding why bigger parties have the motivation to catch up and why the resource advantage (more money, expertise, and media links) is so important in allowing bigger parties to catch up.

We propose that resources and motivation are indeed the crucial building blocks that determine whether the opportunities offered by social media are fulfilled and whether the diffusion process occurs empirically as specified in Gibson and McAllister's multistage model and develop these thoughts further. Some small parties might have an initial advantage in terms of being motivated because of social media's specific characteristics, but the larger parties will be able to dominate some social media as the more expensive opportunities come within reach when such media are used professionally. Let us unpack these mechanisms and arguments. Thereby, we conceptualize the multistage process as running parallel to the generally accepted diffusion of technological innovations (Rogers, 2002) and connect that to the important roles that motivation and resources play, especially concerning the aforementioned social characteristics and opportunities.

Innovations generally tend to be diffused via a bell-curve pattern, with innovators being followed by early adopters and the early majority. Later on the late majority start using the new product too, with (Luddite) laggards joining last, or not at all (Rogers, 1995). For each of these three phases of social media's development, we will formulate expectations.[8]

Early Adoption: Expecting Equalization (for Some)

Once enough early adopters are active on social media (i.e., 10–15% of the population), social media can be expected to become particularly interesting for smaller parties, given the advertisement opportunity. By then there is a substantial electorate or audience active on social media, and because they lag behind electorally, smaller parties can be expected to have more incentives to try new communications and campaigning means. The low start-up costs allow politicians to try out social media. Large parties do not have to fight uphill, so we can expect them to stick to the routines that have proven their value, the same routines that have brought them the powerful position they are in. Here social media's human-contact opportunity might also play a (minor) role. As smaller parties have a greater incentive to convince people, they might be more prone to see the human-contact opportunity, whereas dominant parties might be more focused on the other political-communication arenas and behave somewhat more arrogantly regarding interactions with their large and diverse electorate. Larger parties might even be afraid of attracting negative attention, as they are constantly in the spotlight and an easy target if they start showing and interacting more.

In this early-adoption phase, the notion of motivation and social media's characteristics also suggest that within the group of smaller parties, these new opportunities fit some smaller parties better than others. So far this division has been overlooked. This is probably because of the research design. Indeed Gibson and McAllister (2014) come to their "first equalization, then modernization" conclusion by comparing the two major Australian parties and just one minor party, the postmaterialist Greens, so they cannot distinguish party size from ideology. Vergeer and Hermans (2013) are more explicit about this subject, arguing that ideology matters in terms of motivation. They draw particular attention to populist and postmaterialist parties, the former because of the supposed anonymity online (but this is very unlikely, given that Web 2.0 is rarely anonymous, in contrast to Web 1.0). For the postmaterial parties, the interactive nature of social media would indeed fit their ideology. To start with the latter, postmaterial parties such as the Greens can certainly be expected to be more motivated, because it is exactly their electorate that is often overrepresented among those early adopters. Additionally, presenting their parties as being modern and progressive by using social media fits them and their electorate well. Moreover, not only will their electorate be among the early adopters, but the party cadre and the politicians themselves are very likely to be as well. This implies that even without resources they are capable of going beyond simply being online, as they themselves will build up expertise in order to use social media professionally and effectively. Populists, as nonmainstream parties, can be expected to have a certain degree of motivation too, one that is linked to the defining aspect of a populist ideology: the Manichean conceptualization of society being divided between an evil, corrupted elite and the good, "ordinary" folk (Mudde, 2004). Populist leaders consider themselves the spokesmen or -women of such ordinary people. This association shows some overlap with the human-contact opportunity of social media, and we could therefore expect them to be early adopters too. At the same time, populist parties might especially have the strongest incentive *not* to use social media. Indeed, populist parties are often organized around strong leaders who are the personification of the party and want to keep control over the party as a whole. As a result, populist parties are often highly centralized (see Mudde, 2007; Van Kessel, 2015). From that perspective, social media's advertisement and salon-debate opportunities might limit the control of the leader and deincentivize these parties from using social media. For populist political actors, social media clearly are a mixed blessing, and the overall effect depends on the relative weight of the positive and negative motivation, which in turn might differ per populist party.

In sum, we can formulate the following expectations:

Exp. (1) In the early-adoption phase, equalization takes place, with several smaller parties making use of social media (and benefitting), while the bigger parties do not have the motivation to enter this new political arena.

Exp. (2) Of the smaller parties, postmaterialist parties are particularly well placed, as they combine a very strong motivation with professional expertise as a resource.

Widespread Use: Expecting Normalization for Bigger Parties

When the diffusion of social media continues, a large part of society (the early and late majority) adopts them at least to some extent. This has two important implications. First of all, the public on social media becomes more diverse, and societal entrepreneurs, opinion leaders, news scouts, and so on will be more present on social media from now on. Second, since some time has elapsed during which the minor parties and politicians have been using social media, mainstream parties and politicians will have been introduced to it and will have seen the potential benefits of social media. Considering the notions of resources and motivation, we can expect normalization to take place in this phase: The mainstream parties will now have more incentives (motivation) to enter the social-media stage.

When a majority of the population is present on social media, the motivation of the larger mainstream parties is evident: As their electorate and supporters are on social media, and they themselves are not, but their smaller competitors are, their dominant position is threatened. Yet they lag behind at least some smaller parties in terms of social-media experience and skill. Of course mainstream parties can easily open a social-media account, and the advertising opportunity might also be easy enough to seize. However, making good use of the other opportunities is a matter of trial and error, expertise, and an all-round communication strategy: which kinds of posts are most effective; how to use social media in a professional way; how do you reach a diverse audience that is composed of multiple target groups? Here the resources element enters the picture. Larger parties can buy expertise via social-media consultants, they can push their accounts and posts through ads, and they can invest in professional and continuous webcare to monitor what is going on. In short: Resources enable them to catch up.

In this respect it is important to realize that in the phase of widespread diffusion, *professional* social-media use is more important than it was in the early-adopter phase. After all, in this phase, some smaller parties have already built up experience, and the bar regarding what is considered good use has been raised. For instance, a party or candidate will not look modern in this phase simply by having an account with a few messages. On the contrary, they are more likely to be considered amateurs. Consequently, it is far more difficult for smaller parties to catch up in this phase, as they do not have the resources to buy their way in. More on that below, where we discuss the third phase.

To summarize, our expectations are as follows:

Exp. (3) In the widespread-diffusion phase, normalization will take place to the extent that the larger, mainstream political actors adopt social media at a high speed, as they now have the motivation to do so electorally and the resources to buy expertise.

Exp. (4) If smaller parties did not start using social media early on, because motivation was lacking, for instance, because of the loss of control populist parties might fear, they will not be able to catch up with the other small and mainstream parties, given the now required expertise.

Laggards: Expecting a Shift from Unequalization to Normalization

In the last phase of the diffusion process, the laggards start to use an innovation. As mentioned earlier, this might be an incentive for some smaller parties who were still at best only barely using social media to enter the stage. Such parties probably stayed away from social media for two reasons. The first barrier to using social media is that of the loss of control, which will particularly be an issue for highly centralized parties. The second one entails a mirror argument of why the postmaterial parties will be the frontrunners: Some small parties appeal mainly to electorates who are among the last to adopt social media. When the potential supporters of a party are not on social media, such media are probably not part of the subculture from which the party's politicians are recruited, and there are few potential voters to win or citizens to connect to on social media. The best examples of such parties would be those that are generally adverse to technological innovations, such as ultraorthodox Christians or parties who represent largely computer-illiterate electorates, such as parties focusing mainly on the elderly. In these cases, the first three opportunities offered by social media are not really important, and only the salon-debate opportunity might provide a significant motivation to enter the world of Web 2.0. Consequently, they will have a considerable lag in social-media use, network size, and above all expertise and professionalism. As they have neither the financial resources nor the cadre from which to draw expertise, it can be expected that they will remain behind even if they enter Web 2.0. In other words, the frontrunner parties (both small and mainsteam ones) will remain far ahead of them, and it will take considerable time for them to catch up with respect to social media (but by then other innovations might have entered the front-runners' political communication toolbox, and social media might be a conventional basic tool). In this phase, it becomes clear that social media enable some smaller parties to make inroads in the territories of other smaller parties, while the bigger parties will not be surprised. In sum, while most research focuses on small versus big parties, what seems to matter far more is how some small parties perform compared to *other small parties*.

In a nutshell, we can add one more expectation.

Exp. (5) During the laggard diffusion phase, the most traditional small parties will also start using social media, but because of a lack of resource and expertise they will remain behind, normalizing their position as a smaller party.

Extending the Motivation-Resource-Based Diffusion Model

Above we formulated our motivation-resource-based diffusion model and sketched how the political opportunities of social media can trans-late to the everyday political power balance between (candidates of) political parties ("interparty politics"). We did this in the light of the existing equalization and normalization debate and derived more pre-cise expectations that will be tested in the remainder of this book.

The unique characteristics of social media, however, point toward at least three more avenues along which our understanding of the impact of social media on the power balance in politics could be developed: (1) platform differentiation, (2) political context dependency, and (3) intraparty politics. As these three avenues are largely absent from the equalization-normalization literature, but can be addressed using the logic presented above, we will develop further research questions here, and sketch a picture of what impact social media seems likely to have. Empirically they will also be addressed later on in the book.

Affected by Which Social Media Platform?

While different platforms share the typical characteristics of social media, it is hard to argue that platforms as diverse as Facebook, Google+, Instagram, KIK, Snapchat, Meerkat, and Twitter are exactly the same. Nevertheless, most research focuses only on Twitter (e.g., Jacobs & Spierings, 2014; Kruikemeier, 2014; Peterson, 2012; Vergeer & Hermans, 2013; Vergeer, Hermans & Sams, 2013) or lumps different platforms together (e.g., Hansen & Kosiara-Pedersen, 2014; Koc-Michalska, Gibson, & Vedel, 2014). As the aforementioned social-media platforms have different functionalities, the five typical social-media characteristics will be present to different degrees. As a result, the way in which they might influence the political dynamics can be expected to vary substantially, which leads us to the following additional research subquestion:

> RSQ (1): Is there a connection between the different functionalities of social media platforms, most particularly Facebook and Twitter, and the degree to which they lead to equalization or normalization in the different phases of diffusion?

The relevance of this question can be illustrated by two examples dealing with the two most politically important social media (see chapter 3): Facebook and Twitter. Facebook uses rather complex algorithms that punish "some" or "bad" messages and reward "other" or "good" ones. Additionally it sells advertisement space. Its profiles are not necessarily publicly visible and typically people have to mutually acknowledge each other as connections before they show up on each other's wall.[9] It also allows for more visual and personal content, including the brands, bands, tv programs, and even cities people like. Because of regular changes in the algorithm, no one knows exactly what makes a post go viral or even be seen by a majority of someone's followers or connections. It seems that posts including visual material do better (especially video content uploaded to Facebook directly) and that posts of accounts that have a history of more engagement (likes, shares, and comments) are more often shown on other people's newsfeed via the "Top Stories" function. Facebook is thus a complex personal pegboard or magnet-cluttered refrigerator door. Twitter, however, is open and only allows short messages or soundbites and headlines to be posted, and posting visual material in tweets directly is a relatively new feature. Moreover, it allows for one-directional relationships, and Twitter shows all tweeted messages chronologically, until recently without people's timeline being interrupted by ads. In short, Twitter is more of a personal press agency.

Considering these differences, the smaller parties can be expected to particularly benefit from Twitter's salon-debate opportunity in the first and second diffusion stages as the medium matches journalists' needs, given that its short-message format and "quotability" make it a treasure trove for journalists (Peterson, 2012). In the widespread-use phase, the bigger parties can perceive Twitter as a real threat in particular because some of their smaller competitors use social media to attract the attention of traditional media. However, it is Facebook where we can expect resources (money and expertise) to advantage them most, both due to its more multifaceted and visual set-up as well as its complex algorithms and more developed paid "post-boosting" options. Professional Facebook use seems much more demanding than that of Twitter. In other words, it might be Twitter that pulls in the dominant parties, but Facebook may help them restore their dominant position. Finally, in the laggard phase, it seems easier for the small, resource-deprived parties to catch up on Twitter, both because journalists have an incentive to follow parties and politicians and because Twitter is less demanding and the easiest to use. Given the salon-debate opportunity, these parties can also be expected to enter the Twittersphere before they start investing in and catching up on Facebook.

Exploring the Moderating Impact of Different Political Contexts

So far, the equalization-normalization debate has been taking place in general terms, with the focus mainly on Anglo-Saxon, majoritarian democracies (see Conway, Kenski, & Wang, 2013; Evans, Cordova, & Sipole, 2014; Gibson & McAllister, 2011; Gibson & McAllister, 2014; Jackson & Lilleker, 2009; Katz, Barris, & Jain, 2013; Lassen & Brown, 2011; Lilleker & Jackson, 2010; Parmelee & Bichard, 2012; Peterson, 2012), though this was not exclusively the case (e.g., Spierings & Jacobs, 2014; Vergeer, Hermans, & Sams, 2013). One of the reasons why different studies come to different conclusions may be that the political context shapes the impact of social media. This seems likely, as the opportunities social media offer are closely linked to specific characteristics of the political systems, such as whether politicians are elected on a geographical basis, whether media are partisan or not, and how parties are financed. Our second additional research subquestion inquires how the political context moderates the effect of social media:

RSQ (2): How does political context affect the impact of social media on the power-balance position of parties and politicians?

In formulating the possible answers to this question we will focus on our two counterpoint cases: the Netherlands and the United States. Both countries are distinctly different in their political contexts, particularly the electoral system and party-finance legislation. In addition to these characteristics, the type of political arena (local, national, supranational) is a third element of the political context that may be important.

(1) *Electoral system.* Applied to single-member districts, the American first-past-the-post electoral formula is fairly straightforward: The candidate who receives the most votes in the district gets elected, and that politician then becomes the representative and face of that district. In the Dutch list-proportional system, however, the votes are added up per party list, making MPs representatives of a party instead of being elected with a personal mandate. Smaller parties here have a far higher chance to get people elected and sometimes even enter government.

These characteristics suggest that social media's effect in proportional systems might be somewhat more prone to equalization and in majoritarian systems more to normalization. Indeed, in first-past-the-post systems, smaller parties are typically fringe parties with a small pool of supporters and activists, parties that lack overall professionalism. In proportional systems, smaller parties often are not on the fringe but play a genuine role in politics instead. As a result they have a fairly big pool of supporters and a decent level of professionalism.

The overall emphasis on parties rather than persons in the Netherlands also presents a marked difference with the United States. Indeed, in proportional systems like the Dutch one, the connection between MP and electorate is more abstract and less personal (Suiter, 2015), which might mean that the target-group opportunity of social media can particularly make a bigger difference for small parties in such a system. In proportional systems, small numbers of voters can make the difference and yield an electoral bonus. Single-issue parties genuinely stand a chance to get a few representatives into parliament. Indeed, the only single-issue parties that stand a chance in first-past-the-post systems are ones that are geographically focused, regionalist parties such as the British Scottish Nationalist Party or the Canadian Parti Québécois. This makes a target-group strategy worthwhile, whereas catchall strategies are more effective in majoritarian systems (cf. Vergeer, Hermans, & Sams, 2013). Additionally, the salon-debate opportunity might mean more in proportional systems than in majoritarian systems, because even small parties can have some political power in proportional systems, where they are not constrained to the margins of politics, increasing their newsworthiness for journalists.

(2) *Party-finance legislation.* In the theoretical framework, we already highlighted the importance of financial resources. While the impact of party finances is typically examined in first-past-the-post systems like the United States (Scarrow, 2007), generous/limited public party financing or strict/liberal campaign-finance regulation is not confined to such systems (Van Biezen, 2004). Hence we discuss it as a separate element of the political context. Indeed, liberal campaign-financing rules are often accompanied by limited public financing (and the reverse). As public financing is often based on the number of votes or seats a party gets, smaller parties tend to get some small slice of the cake as well. If most financing is private, this favors parties and politicians that have a higher chance of being successful (Scarrow, 2007). Clearly the electoral system interacts with the party-finance legislation. One could expect though that in contexts where at least some funding is available to smaller parties, equalization might be more likely. However, because we do not know the precise impact of financial resources on the equalization-normalization processes when it comes to social media, its precise impact is yet unclear.

(3) *Type of political arena.* Most social-media research focuses on first-order, national elections (Gibson & McAllister, 2014:2),[10] which by default attract a lot of attention from citizens and media alike. One can thus expect that the added value of social media is relatively limited. In second-order elections such as subnational or supranational ones, this is not the case and the target-group opportunity can particularly help small parties to mobilize voters on their core issue. Additionally, the value of the advertisement opportunity is higher in contexts where voters cannot rely on elaborate media coverage to collect information about candidates and parties. Social media not only show voters the official profile of the candidates and parties but they also provide information about those candidates and parties' activities and expertise. When information is scarce due to a lack of mass-media coverage, this potential may actually matter more, and, given that any party can be present

on social media cheaply, this levels the playing field. The effect may however be different for different political arenas. We focus on the two most extreme ones, the local and supranational (European) arenas.

The *local arena* is often characterized by a lower level of professionalization, and even larger parties have markedly less resources and expertise available for these elections. This may postpone the normalization process. And in this light, it can be crucial whether the local party is a branch of a national one, which has expertise to be shared, or whether it is an independent local party. Not only are the resources lower at the local level, the motivation to take to social media may also be lower. For instance, the target-group opportunity depends on the existence of geographically scattered niche groups. On the one hand, as the geographical area to cover is smaller in the local political arena, the target groups may actually be geographically concentrated or are in any case targetable through a ground campaign. Other forms of campaigning that employ direct in-person communication may thus be more important. On the other hand, big cities may still be conducive to a ground campaign in which social media may serve as a proxy to personal contact, especially if politicians interact with their followers. In sum, depending on the size of the municipality, one can expect the overall social-media adoption process to be slower at the local than at the national level.

This is not the case at the supranational, *European* level. Such elections are also second-order elections and yield less attention from traditional media (De Vreese, 2001), but the geographic logic is evidently different. The financial resources are higher at the European than at the local level. European politicians and parties also have more incentives to use these resources. Members of the European Parliament (MEPs) are often geographically speaking more distant than national MPs (as they reside in Brussels), and MEPs hardly get any media coverage. For them, building a strong social-media network might be more important and beneficial. Social-media opportunities therefore are relevant in this setting: politicians and parties can connect to citizens, contact journalists, and advertise their policy positions and expertise. This would suggest the early-adoption effect to be extra strong. However, one can actually expect both the major and smaller parties to benefit from social media, so it is unclear whether we should expect equalization or normalization. In sum, whereas we can expect social-media adoption to be slower at the local level, we can actually expect it to be faster at the European level.

All in all, there is enough reason to believe that the type of political system, the type of party financing, and the level of the political arena all shape the adoption and impact of social media, but their precise influence is difficult to predict. This is partly because of the multiple potential mechanisms, but the effects across contexts might even be similar, yet grounded in different mechanisms. In sum, before we can build hypotheses, more empirical exploration and theory building is needed.

Exploring the Impact of Social Media on Intraparty Relations

As mentioned earlier, Web 2.0 has created new communication opportunities, stimulating a new campaigning strategy that is "personalised, decentralised and unsupervised" (Vergeer, Hermans, & Sams, 2013:480–481). This clearly echoes the definition of the personalization of politics, which shifts the attention from parties to individual politicians. While the equalization-normalization thesis has mainly been applied to the power balance between parties, our theoretical framework suggests that social media can also lead to shifting power balances within parties. More particularly, it raises questions about whether social media's characteristics compromise the power of the party vis-à-vis individual politicians (Vergeer, Hermans, & Sams, 2013) and which politicians benefit from the personalization: the party leaders, politicians representing certain marginalized social groups, or the lower-ranked politicians in general. The equalization-normalization logic can be extended to these intraparty debates and can help to understand issues of representation (see Celis & Childs, 2008; Celis et al., 2008) and the personalization of politics (see Van Aelst, Schaefer, & Stanyer, 2012). To explore this further we formulated our third research subquestion:

> RSQ (3): How do social media change the power balance of individual politicians within parties vis-à-vis each other and the party organization?

All four opportunities offered by social media point toward a power shift from the party to individual politicians. This might be particularly important in more party-centered systems where individual financial resources are limited. While this expectation is quite straightforward, the question remains whether politicians make use of these opportunities, why they do so (or not), which opportunities are the most promising, and whether or not specific social media are favored.

The most interesting question in terms of equalization and normalization therefore is: Who benefits the most? Given the different opportunities social media offer, we can suggest different transformations in this respect. The "personalization of politics" debate distinguishes a general shift to the politicians and a shift toward the top politicians and party leaders (Jacobs & Spierings, 2015; Van Aelst, Schaefer, & Stanyer, 2011). In addition, the debates about descriptive and substantial representation in politics draw attention to the under- and overrepresentation of certain groups of politicians, such as women and ethnic minorities (Celis et al., 2008; Weldon, 2002). Social media provide an arena to which marginalized political groups can tailor their representative strategies, because the gatekeepers from the old arenas are largely absent (cf. Blumler & Kavanagh, 1999; Enyedi, 2008).

Returning to the opportunities offered by social media, the salon-debate, human-contact, and advertisement opportunities suggest that particularly low-ranked politicians and traditionally underrepresented groups have an incentive to start using social media in order to create name recognition, build their own network, and try to make it into the traditional media (see also Lasorsa, Lewis, & Holton, 2012). Indeed, social media not only present an opportunity to level the playing field for previously disadvantaged parties but also for previously disadvantaged *individual politicians*. Especially Twitter is cheap and easy-to-use, and it may grant access to the regular media, making it highly attractive to such previously disadvantaged politicians. Such candidates may still have fewer resources, but their motivation to use social media is higher.

The target-group and human-contact opportunities fit an identity-based form of personalization best. Establishing oneself as the spokesperson of an ethnic minority or underrepresented group (such as women) becomes easier without gatekeepers, and Facebook, in particular, can facilitate such a strategy. Evans, Cordova, and Sipole (2014), for instance, found that female candidates tweeted significantly more than male candidates in the run-up to the 2012 US House of Representative elections. Part of the motivation of politicians representing marginalized groups might also come from the fact that postmaterialist voters are overrepresented on social media in the early-adoption phase (Conway, Kenski, & Wang, 2013), and postmaterialists tend to be more positively oriented toward identity politics, emancipation, multiculturalism, and affirmative action (Inglehart, 1997). In line with this, women's overall political participation seems to be slightly higher on social media than that of men, whereas ethnic inequalities are actually aggravated online, a phenomenon that has been ascribed to their lower socioeconomic status, Internet access, and degree of postmaterialism

(Quintelier & Theocharis, 2012; Schlozman, Verba, & Brady, 2012; Valenzuela, Park, & Kee, 2009).

The intraparty impact of social media in the early-adoption phase can thus be expected to be similar to the interparty impact: equalization (for some). In the widespread-use phase, normalization can be expected, with the top politicians becoming more dominant again, particularly on Facebook, but to a lesser extent Twitter as well.[11] In the laggard phase we can then expect the final batch of politicians to start using social media, most likely because their parties "stimulate" them to do so.

The Next Step: Theory Testing and Building

In this chapter, we set out to systematize and expand the existing debate on how social media affect existing political power balances. We did so by first outlining what it is that makes social media a unique communication tool: they are unmediated, personal, and interactive, able to go viral, and are cheap and easy to use. Next we discussed what opportunities these features offer politically: advertisement, target-group, human-contact, and salon-debate opportunities. We then briefly discussed the potential dark side of social media, which seems overshadowed by the opportunities. Next, we summarized and diagnosed the equalization-normalization debate, showing that both sides use some underdeveloped argumentation.

Building on the two-stage model of Gibson and McAllister (2014) and a general conception of the diffusion of social media in society (Rogers, 1995), as well as the social-media opportunities formulated earlier, we then theorized the impact of social media and developed a motivation-resource-based diffusion model. Doing so, we also more clearly defined which motivations and resources will play a role in these processes, underwhile refining the existing theoretical models. This framework will be empirically tested in chapter 4 using social media data, interviews, and secondary material. Based on our assessment of the equalization-normalization debate, we also formulated three more exploratory research subquestions, and we provided some first explanations for how these scope conditions regarding the political context would affect overall equalization and normalization processes. The first subquestion dealt with the differential impact of different platforms. This question will be empirically assessed throughout chapters 3 to 7, and chapter 8 (the conclusion) will bring those results together. The second subquestion explored the moderating impact of the political context, highlighting the role of the electoral system and party financing.

This question will be assessed by comparisons throughout the book between the empirical analyses reported here and existing research both on the United States and on European countries. The second element of this subquestion was the political arena one focuses on, and we specifically explored the differences between the local and supranational levels. This subquestion will be examined in chapter 6, which will analyze several local elections in the Netherlands and compare these to the earlier empirical results in chapters 4 and 5. Our third and last exploratory subquestion directed our attention to the intraparty power balance and how this balance is affected by the introduction of social media. We focused specifically on previously disadvantaged politicians and the representation of women and ethnic minorities. This question will be addressed in chapter 5.

CHAPTER 3

Social Media in Politics: The Netherlands from a Comparative Perspective

Introduction

In chapter 1, we already provided some information about the Dutch political system and the roles that social media play in Dutch politics. In this chapter, we will present a more complete picture of Dutch politics and of the way social media found their way into it. Aside from sketching out the Netherlands' political system, this chapter also serves as a descriptive case study of social media and its use in general in Dutch politics. In short, this chapter will study the country level, whereas the later chapters focus on the party level (chapter 4), the candidate level (chapter 5), and the local and supranational level (chapter 6).

The first sections of this chapter will acquaint the reader with the Dutch case, focusing on the Dutch political and (social) media system and how it compares to other established Western democracies. The subsequent sections will examine whether the four opportunities social media offer actually materialize at the general country level, thereby already putting some of the theoretical assumptions to the test. Below, we combine several types of data: expert interviews, social-media use statistics, as well as more content-driven media analyses. Combining these different types of data enables us to draw a rich picture of social media's adoption as it develops.

The Dutch Political and Media System

Below we lay out the main features of the Netherlands, focusing on the political and electoral system, the political parties and their candidates, and political campaigning and the general (social) media system.[1]

The Netherlands

The Netherlands is a country in the northwestern part of Europe with about 17 million inhabitants on an area of 42,000 square kilometers (18% of which is water). Its population is composed of about 75–80 percent ethnically Dutch people and about 6 percent EU citizens living in the Netherlands, with the rest largely made up of other ethnic groups with non-Western backgrounds, reflecting both labor migration (Turks, Moroccans) and the Dutch past as a colonizer (Indonesians, Surinamese, and other Caribbean people).

The Electoral System

The Netherlands has been a constitutional monarchy since 1815 and a parliamentary democracy since 1848. In 1917, general suffrage for men was introduced, and the first-past-the-post electoral system was replaced by a proportional one. In 1919, the vote was extended to women, and since 1985 immigrants have been allowed to vote in local elections under certain conditions (Andeweg & Irwin, 2005).

Ever since its introduction in 1917, the Dutch proportional electoral system (see Table 3.1) has remained fairly stable. It is a list system

Table 3.1 Main institutional characteristics of the Dutch political system

Topic	Description
Form of government	Parliamentary system
Parliament layout	Two chambers: Lower house ("Tweede Kamer"): 150 MPs Upper house ("Eerste Kamer"): 75 MPs
Electoral system	List-proportional system; de facto one single district with an electoral threshold of 0.67%
The act of voting	Each voter has one vote. People can only vote for one candidate within a party list
Preference voting[a]	From 1997 onwards; a number of votes equal to 25% of the electoral quota is needed for a candidate to be elected out of the list order.

[a] At the European level this has been 10 percent of the electoral quota since the 2009 elections.

whereby parties present themselves to the voters through a list of candidates. Voters can only cast *one* vote for *one* candidate on *one* of the party lists; they cannot vote for a party, they can only vote for a candidate (see Figure 3.1). All the votes that are cast for the candidates on a party list are added up nationwide and this total number of votes determines the number of seats that this party gets (Andeweg, 2005). No specific registration is necessary for people to be allowed to vote as the government sends all eligible citizens a voting pass. The electoral system is the same for the subnational and European levels. At the subnational level, Dutch citizens vote for 1 of 12 provincial legislative assemblies (39 to 55 seats) and for 1 of the 393 municipal councils (9 to 45 seats).[2] The provincial elections were held in 2011 and 2015, and the local elections in 2010 and 2014 (not concurrently with the parliamentary elections in 2010 and 2014). In 2014 the Dutch also held the elections for the European Parliament, in which they vote for 26 seats, as was the case in 2009.

The Dutch system is highly proportional in terms of the votes-seats ratio, and, given the de facto electoral threshold of 0.67 percent of the votes, it is relatively easy for new parties to enter parliament. As a result, the Dutch parliament houses many smaller parties. Parties receive state funding and parties with more seats in parliament are allowed to list more candidates in the next elections, all of which fosters a multiparty system. In 2010, lower-house seats were allocated to 10 parties, and in 2012 to 11 parties. Nevertheless, the coalition governments that govern the country have historically been dominated by only a handful of parties. Until 1977, all post–World War II governments included at least one of the major Christian parties, and all government until 2015, except two, included the conservative liberals (VVD) or social democrats (PvdA)—the other two included only Christian parties. After 1977, the three major Christian parties merged in the Christian democratic party (CDA), and until 1994 all governments at least included the CDA. Since being founded in 1966, the progressive liberal party, D66, has been a regular coalition party as well. On average these four parties are also the biggest parties (see Table 3.2). The three traditional parties are still the strongest, but D66, the socialists (SP), and the populist radical right (PVV) are challenging them.

Ballot Structure: From Party to Politician

The votes also determine which candidates get the seats. The 10 and 11 parties that obtained parliament seats in 2010 and 2012 listed around

STEMBILJET voor de verkiezing van de leden van het Europees Parlement op donderdag 22 mei 2014

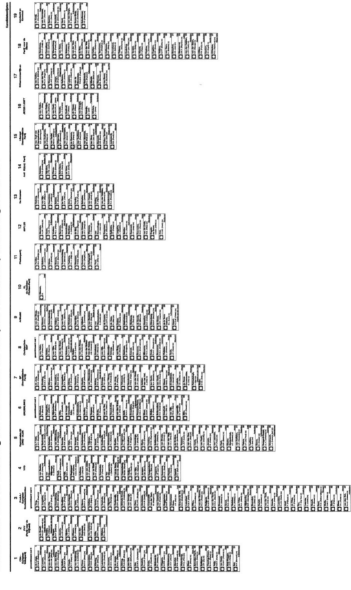

Figure 3.1 A typical Dutch ballot (European Parliament elections 2014).

Source: Kiesraad, personal communication, 2015.

Table 3.2 Election results in most recent elections

	2009 EU	2010 loc.	2010 parl.	2011 prov.	2012 parl.	2014 loc.	2014 EU*	2015 prov.*	Average
VVD—Conservative liberals	11.4	15.7	20.5	19.6	26.6	12.2	12.0	15.9	*16.7*
PvdA—Social democrats	12.1	15.7	19.6	17.3	24.8	10.3	9.4	10.1	*14.9*
CDA—Christian democrats	20.1	14.8	13.6	14.1	8.5	14.4	15.2	14.7	*14.4*
D66—Liberal democrats	11.3	8.2	7.0	8.4	8.0	12.1	15.5	12.5	*10.4*
PVV—Populist radical right**	17.0	0.8	15.5	12.4	10.1	0.6	13.3	11.7	*10.2*
SP—Socialist	7.1	4.1	9.8	10.2	9.7	6.6	9.6	11.6	*8.6*
GL—Greens	8.9	6.7	6.7	6.3	2.3	5.4	7.0	5.4	*6.1*
CU—Center-right orthodox Christian	3.4	3.8	3.2	3.3	3.1	4.1	3.8	4.4	*3.6*
SGP—Ultraorthodox Christian	3.4	1.8	1.7	2.2	1.1	2.0	3.8	3.1	*2.4*
PvdD—Animalp arty***	3.5	NA	1.3	1.9	1.9	0.5	4.2	3.5	*2.1*
50Plus—Seniorsp arty***	NA	NA	NA	2.4	1.9	NA	3.7	3.4	*1.4*

*Combined votes for CU and SGP are proportionally divided in provincial and equally in European election.
**The PVV only participated in two municipalities.
***PvdD and 50Plus did not participate in all elections.
Note: Only national parties are included.
Source: www.kiesraad.nl.

500 candidates all together. After the seats are distributed across the parties, it is determined which candidates will fill these seats. Whenever candidates get at least 25 percent of the electoral quota, they are automatically elected.[3] In the lower house, this threshold typically hovers around 16,000 votes (see Jacobs & Leyenaar, 2011).[4] Roughly speaking, around one-fifth of the seats are filled by candidates who obtained their seat this way. The remaining seats are distributed according to the list order (all the candidates are ranked, cf. Figure 3.1) until all the seats of a party are allocated. In other words, the number of preference votes has priority over the list order. At the same time, only a few candidates who would otherwise have not received a seat purely based on list order obtain one via their preference votes. In 2010, two did; in 2012, only one did.[5] Based on these characteristics, the system has been characterized as "semi-open" or "flexible" (Colomer, 2011:10).

The preference-voting literature has shown a rather robust pattern by which the first woman on the list and the first candidate of a nonmajority ethnic group tend to obtain relatively more preference votes as an expression of the importance voters attach to substantial and descriptive representation (Spierings & Jacobs, 2014; Wauters, Weekers, & Maddens, 2010). With regard to gender and ethnicity, the 2014 share of seats in the Dutch parliament taken by women was 38.7 percent, whereas it was 40.7 percent after the 2010 elections.[6] Moreover, about 10 percent of the members of parliament belong to an ethnic minority, and between 1998 and 2010 roughly 80 percent of the ethnic-minority MPs were women (Mügge & De Jong, 2013).

These data refer to elected politicians, but looking at the characteristics of all the candidates on the party lists is also insightful. Table 3.3 lists data about the candidates of all the parties that obtained at least one seat in the 2010 and 2012 parliamentary elections (see also Celis & Erzeel, 2013). Men and ethnic-majority candidates outnumber women and ethnic-minority candidates respectively. Ethnic-majority men also outnumber ethnic-majority women, but the number of male and female ethnic-minority candidates was more or less equal in the 2010 and 2012 elections. Overall, women and ethnic minorities groups are underrepresented compared to their population shares.

Campaigning

Dutch electoral campaigns largely focus on the party as a whole, with particular attention for the first candidate on a party's list—the political leader—or a select group of candidates (Andeweg & Irwin, 2005). Campaigns are run from the parties' headquarters by a staff that tends

Table 3.3 Ethnic status and sex of the Dutch 2010 and 2012 parliamentary candidates

		Sex		
		Man	Woman	
Ethnicity	Majority	639 (62.4%)	312 (30.5%)	951 (92.9%)
	Minority	39 (3.8%)	34 (3.3%)	73 (7.1%)
		678 (66.2%)	346 (33.8%)	1,024 (100%)

Source: VIRAL;o wnc alculations.

to remain even when the political leadership changes. The lion's share of the campaign budget is thus devoted to the leaders in the form of staff, leaflets, television and radio commercials, posters, gadgets, websites, Internet banners, and meetings and activities. While the political leader is often at the center of the campaigns, posters often do not include a picture of the so-called "list puller," and the party platform gets a great deal of attention as well (Vliegenthart, 2012:144). The parties' main focus is to get their message into the media. Grassroots campaigning and canvassing only play a limited role in the Netherlands (Andeweg & Irwin, 2005:92–97).

Some (lower-ranked) candidates do set up a personal campaign, but they only receive (very) limited resources from the party to do so and even fewer put up considerable amounts of money themselves— though individual campaigning is relatively widespread in the candidate selection phase (in order to obtain a higher position on the list). Once the lists are finalized, it is not that common for candidates to strongly campaign for themselves, but it does happen.[7] The main tools individual candidates have at their disposal are flyers, posters, and personal websites and social media. The latter have become more and more important in this respect, particularly because they are cheap and relatively easy to use (and thereby do not require support from the party).[8]

Media

The importance of television draws our attention to the overall media system in the Netherlands. Like most Northwestern European countries, it is classified by Hallin and Mancini (2004) as a Democratic-Corporatist media system with a high degree of political parallelism, high levels of state intervention to protect the freedom of the press, a strong public broadcast system, and high levels of journalistic professionalism. It is different from Anglo-Saxon countries, such as Australia, the United Kingdom, and the United States, in the sense that commercial media outlets are present but less developed. Furthermore, the Dutch state plays a far more important role in securing press freedom. It should be stressed though that these differences with the Anglo-Saxon countries are fading (Hallin & Mancini, 2004).

Internet access is a necessary condition for using social media. Just like in other West European countries, the Dutch Internet penetration rate is fairly high: on December 31, 2011, it was 89.5 percent, and two years later 94.0 percent. In the EU, only Denmark (94.6%) and Sweden

(94.8%) have a higher penetration rate. In the United States, the penetration rate was 86.9 percent in 2014.[9]

Over the years, and similarly to the United States, the Netherlands has always been one of the frontrunners when it comes to social media.[10] Zooming in on specific social media, three platforms were dominant around June 2010 (when the parliamentary elections were being held): Hyves had roughly 8 million users, Facebook 4.5 million, and Twitter 1.7 to 2.5 million. The three were respectively ranked fourth, sixth, and thirteenth on the list of the most visited general websites in 2010 (Oosterveer, 2012; Vergeer, Hermans, & Sams, 2011). By 2012, Twitter had expanded to 3.3 million users (Oosterveer, 2013), and in same year, the number of Facebook accounts was reported to be between 7.5 and 9 million. The number of Hyves users had by that time however dropped to a mere 1.2 million users (Oosterveer, 2013). The Dutch are also very active on social media. In 2013, 65 percent of the Dutch population reported to have used social media in the last three months, placing it in the top position in the EU, with the UK coming second (57%), and Sweden third (54%). Among 16–24 year olds, a staggering 97 percent of Dutch youngsters use social media daily, again more than in any other EU country (Woollaston, 2013). Moreover, Dugan (2012) reports that the Dutch have the highest percentage of active Twitter accounts, followed by Japan, Spain, the United States, and Indonesia. Despite being from such a small country, Dutch topics regularly become worldwide trending topics on Twitter, and the Dutch government crisis of February 2010 for instance became the most tweeted about topic on Twitter globally (Vergeer, Hermans, & Sams, 2011:479).

The Netherlands in Comparative Perspective

Chapter 1 already highlighted that the political and media system of the Netherlands is very different from the one in the United States: the electoral system is far more proportional, the number of parties is far higher, and the position of public broadcast services is far stronger. In Table 3.4, we extend this comparison and also include two additional Anglo-Saxon countries often studied by political-campaign scholars, namely Australia and the United Kingdom, as well as several large- (Germany and France) and medium-sized continental European countries (Denmark, Sweden, Belgium, and Austria). As the table highlights, the Dutch political system is representative for medium-size countries in continental Europe: All have a list-proportional electoral

Table 3.4 The Netherlands in comparative perspective

	Country	Electoral system	Ballot structure	Effective number of parties	Media system	Population size	Internet penetration (%)	Facebook penetration (%)
Main case	The Netherlands	List-proportional	Semi-open	5.70 (2012)	Democratic Corporatist	16,877,351	95.7	44.8*
Anglo-Saxon Countries	USA	First-past-the-post	NA	1.96 (2014)	Liberal	318,892,103	86.9	51.4
	United Kingdom	First-past-the-post	NA	2.57 (2010)	Liberal	63,742,977	89.8	51.7
	Australia	First-past-the-post	NA	3.23 (2013)	Liberal	22,507,617	94.1	52.5
Continental Europe (large)	Germany	Mixed proportional	NA	3.51 (2013)	Democratic Corporatist	80,996,685	88.6	31.3
	France	Two-round run-off	NA	2.83 (2012)	Polarized Pluralist	66,259,012	83.3	38.7
Continental Europe (medium size)	Sweden	List-proportional	Semi-open	4.99 (2014)	Democratic Corporatist	9,723,809	94.8	50.9
	Denmark	List-proportional	Semi-open	5.61 (2011)	Democratic Corporatist	5,569,077	97.3	54.6
	Belgium	List-proportional	Semi-open	7.82 (2014)	Democratic Corporatist	10,449,361	90.4	47.1
	Austria	List-proportional	Semi-open	4.59 (2013)	Democratic Corporatist	8,223,062	86.8	35.5

*The relatively "low" Facebook penetration rate is due to the fact that these data refer to 2012, when Facebook was still overtaking Hyves.

Sources: Electoral system: Gallagher and Mitchell, 2005; Ballot structure: Colomer, 2011; Media system: Hallin and Mancini 2004; Effective number of parties: Gallagher, 2015; Population size (2014 estimate), Internet penetration (2014), and Facebook usage (December 31, 2012): http://www.internetworldstats.com. Unfortunately no detailed data about Twitter use are available.

system with an open-ballot structure, show multiparty systems, and have democratic-corporatist media systems. The Anglo-Saxon countries are clearly different on these fronts, but regarding the Internet and Facebook penetration rates the differences are fairly limited.

In the remainder of the book, we will place the findings of our Dutch case in a comparative perspective by showing what scholars have found for other countries, like those listed in Table 3.4. This will not always be possible to do in great detail as data on the Dutch case are more abundant than that on any of the other countries (with the exception of the United States), but a fair amount of data and studies are available.

Social Media in Dutch Politics

After sketching the context, it is now time to turn to the use of social media in the Netherlands. How have social media been incorporated in the political system in the Netherlands? How are they perceived and used? In the following sections we adopt a helicopter view, looking at the ways social media are embedded in politics. We particularly study whether the four opportunities are taken up. We will zoom in on Twitter and Facebook, and to some extent on Hyves—the Dutch Facebook equivalent/predecessor. The first two social media are by far the most dominant among politicians in the Netherlands, as well as in many other Western democracies. For instance, looking at the two biggest Dutch parties in 2015, the conservative liberals only refer to Facebook, Twitter, and Google+ on their homepage, and while the social-democrats have buttons on their homepage linking to their accounts on Twitter, Facebook, Google+, Flickr, Youtube, Instagram, Storify, and LinkedIn, the web pages of their politicians only provide direct feeds of their Facebook and Twitter accounts. When we interviewed social-media managers and politicians from the two parties about what they did in the 2012 general election campaign, the answers were very clear. The conservative liberals' social-media manager stated:

> *We have been in touch with Google. Google Netherlands approached us and lent us a Hangout set, including cameras. In the end we did not use it. Why not? Our campaigns are very short and intense, so you always have to think about how this will improve the number of votes you might receive. (. . .) we are not going to invest a lot of time and energy in it, if our voters are not there yet.* (VVD, social media manager, 2013)

The social-media manager of the social democrats told us:

> *We wanted everything to be up and running for the major campaign...so that we could try out some fun stuff....Google Hangouts...I cannot call it a success, but it was mainly a matter of image, that the PvdA was a modern party...regarding the outreach, Google+, I don't buy it. It is not something I will devote a lot of energy to.* (PvdA, social media manager, 2013)

A tech-savvy politician from the social democrats confirmed this:

> *The PvdA has a very strong social-media department, which has very consciously chosen Facebook and Twitter. We also experimented with Google Hangouts, and I participated in that. It was no success, I think, but okay, we experiment a lot. The social media that have proven themselves, Twitter and Facebook, are used extensively.* (Pvda, MP, 2013)

In other words, the politicians and professionals did not believe in other platforms (yet). Even currently, the numbers stack up against the other social media platforms: Twitter and Facebook are the most popular, with over 20,000 up to 70,000 followers or likes for these two parties, whereas the other platforms are generally far behind.[11] For the 2010–2014 time span, with general elections in 2010 and 2012, Twitter and Facebook were clearly the most important social media.[12] So how did the Dutch parties use these two social media?

Advertisement Opportunity: Having a Social Media Account

Table 3.5 displays the number of candidates and parties that had accounts on Twitter and Hyves/Facebook as of 2010. The high activity rate of both politicians and parties illustrates that social media have become very popular among politicians. This is in line with the characteristic of social media that they are cheap and easy to use. Zooming in on the diffusion of social media shows us the quick development of social media from a relative rarity to a political must. In 2010, just before the general elections, a third of the candidates used Twitter during the campaign. However, in two years time, Twitter had proven itself and the activity rate *doubled* to 76 percent, even though only 20 percent of the Dutch citizens was on Twitter at the time (Oosterveer, 2013). As the successor of Hyves, Facebook became popular too and its political usage rose to 72 percent in 2012.

Table 3.5 Social-media diffusion among Dutch parties and politicians

2010 General election			2012 General election				2015 Lower House^c	
Twitter	Hyves	At least one type of account	Twitter	Facebook	Hyves^a	At least one type of account	Twitter	Facebook
All candidates (n = 493)			All candidates (n = 531)				NA	
34%	46%	58%	76%	72%	27%	88%	NA	NA
Elected politicians (n = 150)			Elected politicians (n = 150)				Elected politicians (n = 150)	
45%	48%	63%	86%	80%	31%	99%	95%	NA
Parties (n = 10)			Parties (n = 11)				Parties (n = 11)^b	
10	6 (Facebook)	—	11	10	—	—	11	10

[a] Hyves was dissolved in 2014 as the similar Facebook became ever more dominant in the Netherlands.
[b] Seven politicians went independent between 2012 and 2015, organizing themselves in five new parliamentary groups of one or two MPs each, these are excluded from the numbers.
[c] Moments of measurement: 2010 day before election; 2012 day before election; 2015: March 2nd.
Sources: Own calculations; see also www.RU.nl/VIRAL; Spierings and Jacobs (2014).

Our interview with the social-media manager of the conservative liberals further underscores that candidates use these accounts politically, not just for personal matters:

> *"Among our candidates, quite a lot have their own accounts (…) combining personal and professional usage (…) During the campaign they accepted every friend request and made some political statements besides posting personal photos."* (VVD, social media manager, 2013)

The importance of being on social media seems even more pronounced among those who have already been elected—the parliamentarians. Their activity rates developed in the same way, but they are higher. In 2015, Twitter was used by no less than 95 percent of the MPs. More generally, 63 percent of the parliamentarians had an account on least one social-media platform in 2010, which rose to 99 percent in 2012.

The enthusiasm about social media is not restricted to candidates and MPs. All parties in parliament had a Twitter account in 2010 and 2012.[13] Facebook followed a bit later: Only 6 parties had a party account in 2010, but by 2012 all but one of the parties did.[14] Whereas having a social media account in 2010 made parties and politicians look modern, by 2012 parties *not* having an account risked looking old fashioned.

Overall, the first politicians and parties started to experiment with social media between 2007 and 2009 (early adoption), with the 2010 parliamentary elections being the first in which social media really started to play a role in the campaigning of parties and candidates. By the 2012 elections, the usage of Twitter and Facebook was widespread (late majority), and by 2014–2015, most of the laggards were present on social media. Clearly Twitter was introduced earlier than Facebook, but both are widely used now. Also, we saw the first medium disappear (Hyves) and indications of a new medium becoming important (Google+; see chapter 4). Social media seem to be here to stay in one form or another.

Advertisement Opportunity: Providing Content

The advertisement opportunity of social media obviously requires that parties and politicians do more than just create an account. At the party level this is clearly not the case in 2015, as the parties that have opened an account also use it, and party professionals typically make sure updates are sent out regularly. Indeed all of the nine parties we interviewed had professionals that were responsible for providing social-media content.[15]

The personal accounts of politicians are more sensitive to inactivity. Illustrative is an account opened by the populist MP Dion Graus (PVV) two months before the 2012 elections. Only one tweet has been posted so far:

"Finally decided to join Twitter. #pvv #12092012"

Nevertheless, the account has 188 followers, including other MPs and the parliamentary committee of which Graus is a member.

A newer development is that some tech-savvy politicians close their account. For instance, the former Minister of Foreign Affairs and now European Commissioner Frans Timmermans (PvdA) decided to close down his Facebook account on November 2012. He was known for his highly (inter)active and personal use of Facebook, but after aggressive reactions to his position on the Israel-Palestine conflict, he warned his followers and in the end closed down the account. About a month later, however, he returned to Facebook with a new account and actively propagated that diplomats should be more active on social media (De Valk, 2012). More recently, the political leader of the same party (Diederik Samsom) went from being one of the most actively tweeting MPs to completely stopping in May 2014. This was headline news in many newspapers and on the websites of the TV news broadcasts. After being sick of all the online rants, Samsom took some time off.

Not posting messages, or only showing one's social-media incompetence, might be funny or damaging, but such social-media "zombies" seem to be rare, as our data about social-media activity during campaigns reveal. Among the 168 candidates with an account in 2010, only 18 (11%) did not post a single tweet during the campaign period (roughly six weeks). In 2012, the number of Twitter zombies was 27 of 401 candidates (7%) during a shorter period (four weeks).[16] Focusing on the degree of activity here, we can see a sharp increase in the number of posts over the years. On Twitter we have data for both the 2010 and 2012 campaigning periods. The average number of posts in 2010 was 3.6 tweets a day per candidate; two years later this was 9.7. This increase is partly due to a few extremely active candidates in 2012: The most actively tweeting candidates in the 2010 campaign posted over 20 tweets daily, while in 2012 this was more than 160 a day; that is about once every 10 minutes, four weeks in a row. The five most tweeting candidates in 2012 actually produced more tweets a day than all 168 tweeting 2010 candidates together did a day. One way to correct for these highly active candidates is to look at the median; then the rise in activity is more modest, but still very present: a 40 percent increase in the number of daily tweets (1.9 to 2.7). Again, experience and political professionalism seem to be important. Among the 75 candidates that were on a list and on Twitter in both elections, the median number of tweets a day is 5.5, and the average 14.9. Both figures are considerably higher than the overall numbers.

The parties' accounts show that most parties update their posts regularly, but some parties only allow approved followers to read their tweets or they have not tweeted since the 2012 elections. The ten publicly available party accounts on Twitter have posted an average number of 83 tweets per month since they signed up. Given the overall increased activity over the years, this implies that on average the parties tweeted several times a day. In sum, Dutch politicians are quite active in sending information to their network.

Advertisement Opportunity: Having an Audience

Politicians and parties are thus very enthusiastic about social media, but is this enthusiasm mirrored by the public? It seems that the variation between candidates is very high, mainly because some politicians are extraordinarily popular (see Box 3.1). In 2012, the top five twittering

Box 3.1 Followers top five

Follower Top Five on Twitter in 2012		Friends Top Five on Facebook 2012	
Arjan El Fassed (GL)	266,000	Emile Roemer (SP)*	15,000
Geert Wilders (PVV)*	209,000	Geert Wilders (PVV)*	13,000
Alexander Pechtold (D66)*	163,000	Diederik Samsom (PvdA)*	6,000
Diederik Samsom (PvdA)*	71,000	Marianne Thieme (PvdD)*	6,000
Jolande Sap (GL)*	53,000	Frans Timmermans (PvdA)	5,000

*Listp ullera tt het ime.

candidates consisted of four list pullers and one MP specializing in human rights issues who focused on the Israeli-Palestinian conflict. On Facebook, the numbers were considerably lower. The top candidates in 2012 had between 5,000 and 15,000 friends and consisted of four list pullers and the abovementioned Frans Timmermans. Even among these top-five lists, the differences between numbers of friends and likes are huge, but clearly there are politicians with substantial audiences. Moreover, by March 2015, politicians such as Geert Wilders, Alexander Pechtold, and Diederik Samsom, respectively, had 415,000, 347,000, and 157,000 Twitter followers. All had doubled their follower base, and, given the estimated 3.3 million Twitter accounts in the Netherlands in 2014 (Oosterveer, 2014), these are substantial parts of the population, even if they include some non-Dutch followers.

At the other end of the spectrum, 12 out of 168 (2010) and 51 out of 401 (2012) tweeting candidates had fewer than 100 followers. The same holds for 94 out of 224 candidates on Hyves (2010) and 109 out of 348[17] on Facebook (2012). The average number of followers/friends/likes is modest: 2,000 on Hyves in 2010, 600 for Facebook in 2012, and 5,000 on Twitter in 2010 and 4,000 on Twitter in 2012. That declining average on Twitter is due to *new* candidates entering the political and social-media arena. Indeed, the 75 candidates who were on a party list and had a Twitter account in both 2010 and 2012 saw their average number of followers quadrupled from 1,800 to 7,800. In other words, even lower-ranked politicians have managed to build considerable networks on social media by being around for several years. As also indicated by the social-media manager of the conservative liberals, building a follower base is a "rather slow process" (VVD, social media manager, 2013). Creating an account right before the campaign starts will likely not do the trick.[18]

The most popular candidates have more followers on Twitter than their respective parties, whereas the parties have more friends or likes on Facebook than the candidates. On March 2015, the number of party followers on Twitter ranged from 500 to 77,000, with an average of about 27,000. On Facebook, the average was 22,000 with a range of 3,000 to 49,000, while it was on average 10,000 in September 2012, ranging from 500 to 37,000 (Smith, 2012).

Overall, politicians and political parties seem clearly more enthusiastic about using social media than the public is about following or interacting with them. This most strongly holds for Facebook: While both Facebook and Twitter were almost equally popular among politicians and parties in 2012, the follower base on Twitter is considerably larger.

Advertisement Opportunity: Professional Usage

Politicians and parties' use of social media is just a first step in effective social-media communication, but it is certainly not sufficient in times of professionalization,[19] and we see that parties realize this. Several social-media campaigners mentioned they first started to make manuals and guidelines for social-media campaigning around and after 2012. These were sent to politicians and the local party branches (CU, social media manager, 2014; D66, social media webcare, 2014). For instance, in 2014, one party created a document with ten Facebook do's, including screenshot explanations (see Box 3.2).

While the existence of such documents is revealing in itself—and seems to be widespread, as all the online media managers we interviewed confirmed the existence of some such informal or formal guidelines—they tell us little about the implementation of these guidelines.

Box 3.2 Party headquarters' social media advice

Do's:

- Explicitly reserve some budget to buy Facebook ads.
- Summarize your position in one shareable picture.
- Use the [party]'s "profile picture"
- Change your cover picture regularly
- Start conversations (interact with people)

Source: Personal interview social media manager.

To examine the professional use of social media we analyzed the profiles of all the current Dutch MPs (December 2014).[20] A closer look at these profiles sheds some light on how they present themselves, providing an indication of the degree to which politicians invest in their account. We summarized several professionalism indicators in Figure 3.2.[21] Of the 145 MPs on Twitter, almost all have a picture of themselves as their profile picture. Over 40 percent of the MPs on Twitter have a reasonably professional picture, but 10 percent have uploaded a fairly amateurish one. Furthermore, only about 50 percent of the politicians made use of the option to add a background or banner picture to their profile and less than 10 percent uses their profile to show a bit more of themselves, such as hobbies, family life, or personal experiences. Lastly, only two candidates used their profile to highlight their sex, ethnicity, or religion explicitly. In terms of party affiliation, almost all candidates mention in their profile that they are an MP and which party they are affiliated with. However, fewer than one out of five profiles make that affiliation visible (in the profile picture) for the casual visitor. Also, relatively few candidates add links to the general party website, preferring to link to their own page on the party website or their personal websites. Indeed, it seems that the candidates use social media in a personalized way, prioritizing their own work and agenda. As mentioned earlier though, this should not be interpreted as an increased emphasis on the private lives of politicians (see also Kruikemeier, 2014).[22]

Moving from politicians to parties, the latter's accounts also show some surprising range in their professionalism. All parties have their logo

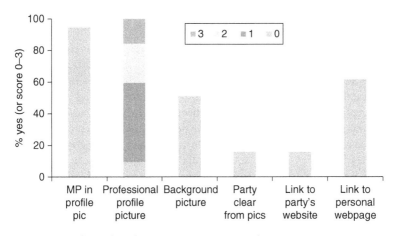

Figure 3.2 ProfessionalismofM PT wittera ccountsi n2 014.

(with or without name) as their profile picture, and four of the eleven included a slogan. Regarding the background picture (see chapter 4):

- Three parties use the standard blue background,
- Two changed the color to a color matching their party logo,
- One uses its name again and added a slogan on the background, and
- Five use a proper picture.[23]

On Facebook, all parties use the same profile picture as on Twitter, with one exception (where the slogan is dropped). This time, nine out of ten parties have a proper background picture.

Do parties use these accounts in a way that fits the medium? One general observation regarding Facebook is that in the early adoption phase (2010–2011) parties had the tendency to publish press releases and include lots of text. By 2015, however, they all posted short messages with large pictures and links to people and websites. This is indeed in line with the criteria that the Facebook algorithm uses to reward certain messages over others (PvdA, social media consultant, 2014). The parties' Facebook use has shifted from a duplicate of the website or press releases to a medium that has a modus operandi of its own. Regarding Twitter, the realization that each medium is unique meant that the parties virtually stopped using the party accounts for sending out messages but instead started to consider it to be a "webcare" tool (cf. interactivity opportunity).

Human Contact Opportunity: Interactivity

Using Twitter as a webcare tool already indicates that for this medium the human-contact opportunity may well be more important than the mere advertisement opportunity. Indeed, as we indicated in chapter 2, interactivity is one of the distinguishing features of social media. The concept of "webcare" seems to have been implemented across all parties, and many parties have made it their policy to at least reply as quickly as possible to citizens' questions. During campaigning periods, several parties have even created social-media teams of volunteers, interns, or even paid employees to ensure questions are answered quickly. One campaign manager voiced this very clearly:

Twitter is a reciprocal medium. When you post something, you also have to answer questions and genuinely interact with people. That is a necessary condition for building a relationship [with followers]. (VVD, social media manager, 2013)[24]

In order to examine the interactivity of the parties more thoroughly, we focus on the Facebook posts of the ten parties that had a Facebook account. In 2012, the parties got an average of about 8,500 to 9,000 interactions (comments + likes) in the last month before the elections, which is an average of 160 comments and likes per post.[25] This does not sound too bad, and, using the IPM score,[26] eight out of ten parties have per-post effectiveness scores that "most brands would consider a finger-licking sensation."[27] Given that likes, shares, and comments (or more broadly "engagement") influence how often posts appear on the timelines of friends of the people who like a party page, this is indeed impressive.

Regarding individual politicians, it is clear that politicians who hardly post messages cannot have interacted much. Focusing again on the 75 candidates who were on the party lists in both 2010 and 2012 and who had a Twitter account in both elections, we see that in 2010, these candidates posted on average three tweets a day, with only four candidates posting ten messages (or more) a day. One can hardly speak of much interaction then. In 2012, these 75 candidates' use had risen to on average 15 messages a day, with 21 candidates posting 10 or more tweets a day. It is safe to say that if any considerable interaction took place, it must have been in the widespread-usage phase from 2012 onwards.[28]

However, the degree of interactivity considerably varies within the group of politicians. For instance, one of the politicians most famous for his supposed Twitter effectiveness—Geert Wilders (PVV)—only tweets and retweets, but never responding to questions. At the other extreme we find politicians who genuinely use social media to obtain information and build connections. For instance, an MP of the progressive liberals (D66) gave us several examples of how she used Twitter to get in touch with individual citizens (see Box 3.3). Similarly, Vera Bergkamp (D66) regularly posts messages on Facebook and Twitter asking for input (see, for instance, Figure 3.3—a similar message was posted on Facebook).

Target-Group Opportunity: Building a Unique Image

The Vera Bergkamp example touches on another opportunity offered by social media: targeting specific niche groups, either in terms of topic or of representing identities. In the case of Vera Bergkamp, before

Box 3.3 A progressive liberal MP (D66) talks about connecting to people using Twitter

"Well, I believe that when people ask explicit questions, I want to respond. This is not always possible, and sometimes you simply cannot answer a question in 140 characters. But those are moments, I mean, when they asked me ... [thinks] I read a comment from someone I follow [on Twitter], a physician, who said that parliament was to debate palliative care. He basically claimed something like, people watch out, they [MPs] have no clue whatsoever, but they are going to talk about it. Well—regardless of the fact that MPs are not physicians and thus have little say in that, only in neighboring questions and guidelines—I did respond, saying that I would like to know more. With people like that I will make an appointment. And things like that, they do start on Twitter. I also did a house call once like that, because that person, a mother whose children were sick, was very worried. You can always just go to the interest groups and organizations that invite you, but I think that direct contact with individuals is also especially important."

MP Vera Bergkamp asks for input on how to increase youngsters' sexual defensibility and adds @-mentions of the main sexual diversity organizations on April 7th 2015.

Figure 3.3 MPB ergkampp ostingi nteractively.

becoming an MP she was the—highly visible—chair of the main Dutch LGBT interest group (COC), and she still connects with that group in terms of identity and policies through social media. Similarly, the social-democrat MP we interviewed aims to represent the self-employed and was very explicit about one of the core advantages of social media in this respect:

"Many candidates are regional based (. . .). My constituency, the self-employed, are everywhere. I cannot be everywhere. But typically self-employed people are all online. For me [using social media] was very obvious, because of my target group." (Pvda, MP, 2013)

At the same time, if we look at the data on the Twitter profiles we used to assess the professional use of social media, we hardly find any explicit issue or identity profiling. In terms of ethnicity and gender, only one of the candidates seemed to explicitly highlight their status as belonging to an ethnic minority by highlighting her "Moluccan roots" in the place-of-residence field (see also chapter 5). No direct indications of religious identity were found; only two MPs included a picture referring to LGBT identities, of which whom also identifies as being gay himself;[29] and only five MPs explicitly profiled themselves as being from a specific and "peripheral" province.[30]

The interviews with social-media campaign managers also indicated that there is little active policy to divide topics or appoint MPs to represent one group in particular. One party advises its politicians to become issue owners on a topic (the conservative liberal VVD) and another party has included "be outspokenly Christian" in their guidelines (CU). These seem to be the exception rather than the rule though. Overall, the target-group opportunity seems to be used to a rather limited extent. Of course we did not analyze the content of all the posts, and MPs may well be focusing on certain topics there. However, if this is the case, it is most likely driven by labor division (i.e., being spokesperson on a certain topic) rather than by identifying with a certain group or wanting to represent that group.

Salon-Debate Opportunity: Connecting to Traditional Media

As discussed in chapter 2, the enthusiasm to start using social media, and Twitter in particular, might not only be to connect with voters directly but also to make use of the salon-debate opportunity. This can have positive and negative effects. The social-media manager of the conservative liberals was clear about his concerns:

Journalists have their Twitter lists ready, they are waiting (. . .) for a [politician] to make a mistake. That is a nerve-racking process for us. We must talk [to our politicians] and raise awareness (. . .) But I have to say that there is, compared to 2010, more general knowledge [on how social media work] already. (VVD, social media manager, 2013)

But as he also indicates, the opportunities have become more important as well, and he particularly adjusted his strategy to the salon-debate opportunity:

This is what we advise our politicians: Select a topic and make sure you become an authority in that area. Make sure that that makes people start to follow you. (VVD, social media manager, 2013)

A campaigner of the Christian party ChristenUnie actually used this opportunity to entice people to use social media:

If I tell them [politicians], "Look, you sent this tweet out this long ago and now it is on Nu.nl [a major Dutch news website]," their ego grows considerably. (PvdA, social media consultant, 2014)

And the progressive liberal MP concurred:

I notice that journalists pick up discussions on Twitter really quickly. Then it features in the national news coverage, and it becomes a hype, while this might not have been the case otherwise [without Twitter]. (D66, MP, 2013)

Twitter and Facebook indeed are important information sources for journalists. A survey among 200 journalists shows that over 50 percent of them—and among the younger cohorts more that 85 percent—uses Twitter as a source of inspiration for stories (Weber Shandwick, 2014:11). For 2010, we also have data on a selection of 15 journalists from leading newspaper, radio, television, and Internet news platforms,[31] and which candidates they followed. Fourteen of the 168 candidates on Twitter were followed by ten or more of these journalists. Each of these candidates was well known. At the same time, 60 candidates were not followed by *any* of the journalists. Still, this means that 108 candidates did forge some connection with the media through Twitter. Among these 108 were many low-ranked and unknown politicians: Their average list position was twenty-first.

Being followed by journalists seems to have become even more important in the last few years. Figure 3.4 shows the number of newspaper articles that mention Twitter or Facebook and the Dutch word for parliamentarian or parliament. This gives a rough indication of the extent to which social-media messages are finding their way into the news. Taking into account that news coverage is lower in June (summer holidays) and particularly high around salient elections (2012 and the

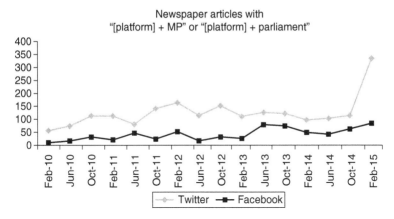

Figure 3.4 Tweets and Facebook posts in the news.
Source: Lexis Nexis; own calculations.

provincial elections of 2015), we can notice an slow but steady increase: During the early-adoption phase, the number of articles for Twitter and Facebook together was around 50, but that number rose to around 200 by the end of 2014.

Facebook as a source for and topic of political news has grown particularly popular since the first half of 2013. Twitter, with its short sound bites, was popular much earlier and has remained the most mentioned of the two. The *Nederlands Dagblad* even has a daily item called "Tweet of the day," and by 2015 politicians' tweets are embedded in the articles on the newspaper's webpages.

Based on the newspaper data, we can therefore say that social media are definitely a way to get into the "old" media. The interviews we carried out suggest that the parties and politicians are aware of this and that they try to make use of this opportunity. Indeed, in line with what, for instance, Broersma and Graham (2012) have concluded about the Netherlands, tweets and Facebook posts seem to be an important source of information for journalists of traditional news media.

Dutch Social Media Use in Comparative Perspective

One important question that remains is whether these patterns are only found in the Netherlands. In this section we compare our findings to those of other countries (see also Table 3.4). Overall, only a few studies exist that present a similar multifaceted look on the political diffusion of social media, and as such this section should be read as an exploration rather than a definitive assessment. Nevertheless, as we show below, the general nature of the findings presented above shows that Dutch politics is indeed at the forefront, at least quantitatively.

Dutch Social-Media Use Compared to The United States

How does the Dutch situation compare to the American one? The United States is often seen as the pioneer when it comes to campaigning and communication tools (Gibson & Römmele, 2001). Indeed, the country also seems ahead of most European ones in social media. Evans, Cordova, and Sipole (2014) analyzed candidates' social-media presence in the 2012 elections for the House of Representatives and show that about 70 to 75 percent of the candidates had a Twitter account. Among incumbents this number even rose to almost 90 percent, and among presidential candidates in 2012, this figure was 100 percent (Conway, Kenski, & Wang, 2013). In terms of social-media adoption, the Netherlands and the United States are thus practically on par.

The numbers of tweets are also similar. In the study of presidential candidates in the 2012 primaries, the two most active candidates and campaigns posted about 70 tweets a day, but most daily averages were about 4 to 10 tweets a day, measured over February and March (Conway, Kenski, & Wang, 2013: Fig. 1, Fig. 4), numbers that are very close to the Dutch ones (9.7 per day on average and 120 for the most active cases). Obviously, the number of followers was much higher than those of the Dutch candidates. Obama alone had around 13 million followers, second came Newt Gingrich with 1.4 million, and Mitt Romney was third with 500,000 early May. But considering that the United States has 20 times as many inhabitants, these figures are quite similar to the Dutch ones (though Obama is clearly an exception), placing the US top candidates in a comparative league with the prominent Dutch politicians again (cf. Box 3.1).

While the United States and Netherlands are rather similar in terms of quantitative diffusion, the quality of Twitter use seems to differ.[32] Most particularly, it seems that in the United States the emphasis is more on the personal. In their comprehensive study, Evans, Cordova, and Sipole (2014: Fig. 2) find that almost a third of the candidates' tweets were personal (e.g., family photos, posts about heading to church). Echoing the more personalized political system, American Twitter-politics seems to be more personalized as well. User interaction made up about 15 percent of all communication among the candidates for the House of Representatives, which is double the percentage of 7.4 percent reported by Golbeck, Grimes, and Rogers (2010) on the Twitter behavior of members of Congress in 2009. Among the 2012 presidential candidates, the percentage of messages including at-mentions varied between 7 percent (Ron Paul) and 66 percent (Gary Johnson), and the candidate average was about 38 percent (Conway, Kenski, & Wang, 2013). Earlier Dutch studies find interactivity ranging from 28.5 percent (2010) to 44 percent (2009) (Kruikemeier, 2014; Vergeer, Hermans, & Sams, 2011).

Dutch Social-Media Use Compared to Other Western Established Democracies

For some countries comparable data are available, showing that the other Western countries are some phases of diffusion behind the Netherlands and the United States. In the United Kingdom, where Twitter was already very popular in 2012 (Woollaston, 2013), a study on the 751 candidates from 177 "close-call" constituencies in the 2010 election showed that 41 percent of the candidates had a Facebook account and 33 percent was present on Twitter (Southern, 2015). This seems similar to the figures presented for the Netherlands, but as the author herself also stressed, these constituencies were chosen because the adaption of social media was expected to be high there given the competitiveness of the elections (Southern, 2015:8), a line of reasoning confirmed by other studies (e.g., Evans, Cordova, & Sipole, 2014). Of all UK MPs in mid 2013, 69 percent were on Twitter (Martin, 2013),[33] whereas in the Netherlands that percentage already was 86 percent almost a year earlier. In Australia, 33.8 percent of the 2013 general election candidates had a Twitter account, averaging 3,770 followers. A lot of the candidates had a Facebook page (66.9%), but the average number of friends was low: 525 (Chen, 2015).

Moving to the list-proportional European countries, among a sample of 1,000 Austrian candidates for the national elections of 2013, 50 percent had a Facebook account and 16 percent had signed up on Twitter (Dolezal, 2015). In Belgium, a sample of 8,070 regional, federal, and European parliamentary candidates (2014) shows activity rates of 18 to 29 percent on Twitter (Jacobs et al., 2015). Larsson and Kalsnes (2014:6) found that in 2013 58 percent of the actual Swedish MPs were on Twitter, while a measly 19 percent had a Facebook profile.

In other words, all these other Western countries are one or two phases behind the Netherlands in terms of politicians' presence on social media. Consequently, we can also expect lower degrees of professionalism, and this explains why relatively many early studies focused on Dutch politics (Jacobs & Spierings, 2014; Kruikemeier, 2014; Spierings & Jacobs, 2014; Utz, 2009; Vergeer, Hermans, & Sams, 2011).

Little is known about the other indicators discussed above, but the limited available information seems to confirm the picture sketched earlier. From the Belgian study, for instance, we know that the tweet frequency is on average two to four a day across candidates in the 2014 campaigns (quite similar to the Dutch 2010 situation) (Jacobs et al., 2015). The aforementioned British study on the 2010 election (Southern, 2015) is one of the most comprehensive ones on the political usage of social media in the United Kingdom and shows that, of the candidates on Twitter, 16 percent did not post *any* messages in the week before the elections. On Facebook, this was true of 34 percent. In terms of interactivity, respectively, 22 percent and 54 percent of the candidates on Facebook and Twitter engaged with followers (Southern, 2015: Table 2). In the Belgian study examining the 2014 general elections, the average number of followers ranged between 100 and 200, and even the list pullers did not have many more than 1,500 followers. In general, we not only find lower degrees of "quantitative" diffusion in European democracies but also lower degrees of "qualitative" diffusion (i.e., the quality of posts and followers). Indeed, the average number of posts and interactivity is relatively high in the Netherlands, and the number of "social-media zombies" lower.

Conclusion

In this chapter, we discussed the political context of the Netherlands and showed the results of a general case study on the quantitative and qualitative diffusion of social media in Dutch politics. This exercise is the first of its kind in detail and scope, providing a general background to further test and develop our motivation-resource-based diffusion model. Additionally, and no less importantly, the descriptive macrolevel analyses in this chapter already put some of the general expectations and assumptions about the political usage of social media to the test. These include the pace of diffusion, how and whether the four opportunities materialize, and whether there are differences between platforms. In what follows we will line up our conclusion for each of these three elements and finish with some comparative notes.

Diffusion: Mission Accomplished?

Based on both our above analyses of the general political diffusion of social media and the studies mentioned in the text, we can say that social media were introduced in Dutch politics in 2007; around 2010, the diffusion was moving from the innovator to the early-adopter phase; by 2012, social-media use was widespread; and by 2014/2015, even the laggards were present on social media. In other words, the diffusion seems to be complete. However, our results suggest that there is an important difference between *quantitative* and *qualitative* diffusion. The qualitative diffusion of social media is still very much in development. This brings us to our assessment of the four opportunities outlined in chapter 2.

Assessing the Four Opportunities

Our nationwide assessment of the four opportunities highlights that there are big differences between the four of them.

1. Advertisement opportunity. Even though (or precisely because) the majority of political actors are now on social media, their overall activity did not decline. Politicians and parties are well aware of the need to do more than merely being present—they need to use social media actively. Generally, their activity has increased and the advertisement opportunity is clearly taken seriously.
2. Human-contact opportunity. With the quantitative spread of social media, the qualitative diffusion has also started, but this process is still under way. Parties and politicians acknowledge the interactive features of social media, but politicians seem to use them rather unsystematically. All parties, however, acknowledged the usefulness of Twitter as a "webcare" tool.
3. Target-group opportunity. The more innovative early-adoption politicians and parties acknowledge that social media offer new opportunities but are hesitant to make full use of those. There seems to be some hesitation in particular about showing the person behind the politician, and while some politicians and the occasional party seems to be aware of the target-group potential of social media, these are the exceptions rather than the rules.
4. Salon-debate opportunity. From the start, one of the main attractions of social media has been the opportunity to gain more media coverage. This seems a matter of an excellent match between supply and demand: Politicians want to be covered in the newspapers

or on television, and journalists like to use social media as an easily accessible source of information. Twitter is everyone's favorite in this respect. However, especially lower-ranked politicians might overestimate the news value of their posts a bit; the number of tweets and posts used by journalists is on the rise but is still somewhat modest. However, if their message makes it, they can reach a large audience that they would not have reached without social media. In this sense, it resembles playing the lottery: You have to enter to be able to win.

Different Types of Social Media

During the course of our examination of the four opportunities it became clear that especially the parties (and some politicians) are becoming more and more aware of the different functionalities of Facebook versus Twitter. Facebook seems to be dominated by parties and Twitter by candidates. Particularly Facebook's complexity and, in some cases, its requirement to add people actively seems to prevent politicians from using Facebook to the same degree as they use Twitter. Moreover, politicians seem to view Twitter as the social medium par excellence to get attention from journalists, whereas Facebook is closer to their private lives.

Comparative Perspectives

Yet can these findings be generalized to other Western democracies? All in all, in terms of quantitative diffusion the conclusion is straightforward: The Netherlands is well ahead of other European democracies and on par with (if not slightly ahead of) the United States. As we mentioned earlier, detailed data about social-media use is surprisingly hard to come by, and scholars seem to pick their own paths rather than examine the same topics in the same way (see also Jungherr, 2014). More generally, reviewing the literature, it seems that social-media adoption is higher on average in majoritarian systems, which might be related to the higher degree of competition between individual candidates there (see Evans, Cordova, & Sipole, 2014; Southern, 2015). Yet this makes the quantitative diffusion of social media among Dutch politicians and parties even more extraordinary.[34] As such, and given that the diffusion of social media seems to follow a rather standard pattern of diffusion, it can be expected that the Dutch case will provide useful insights into what is to come in other countries.

The comparative question whether this is due to certain aspects of the political system or culture is beyond the scope of this study, but the benefit of social media being cheap in a country with low campaign budgets and a system relatively open to smaller (e.g., postmaterialist) parties might play an important role here. In this respect, it is noteworthy that politicians seem to use the human-contact opportunity least, which fits the relatively low degree of diffuse personalization (i.e., privatization) in the Netherlands compared to, for instance, the United States. It is precisely these aspects of political inequalities and power balances that we will examine in Part II of this book.

PART II

Changes in the Power Balance

CHAPTER 4

Interparty Relations:
David versus Goliath

Introduction

Dutch politics seems to have embraced social media. But does this hold true for all the parties? Are there differences between parties not only in their more superficial quantitative diffusion but also in their more substantial qualitative diffusion? Do we see signs of normalization, equalization, or both? These are the topics of this chapter on interparty relations. This will be the first test of the *motivation-resource-based diffusion model* we developed in chapter 2. The model predicts there would be partial equalization at first, then normalization, and in the end just a few parties lagging behind. Behind these expectations lies a logic focusing on the parties' resources (financial and human) and motivations (the identity and ideology of the party and its electorate). As the model is new, the empirical analyses in this chapter also present its first systematic test. In this sense the chapter has goals that move beyond testing theory into the territory of generating and refining theory. By closely examining the Dutch case we can further refine *how* motivation and resources affect social-media use.

In what follows, we start by briefly explaining how we conducted the analyses, then we describe the Dutch diffusion process at the party level, in terms of both the quantitative and qualitative diffusion among parties and candidates in the next sections. Once we have described what happened, we move on to applying the model, first zooming in on the parties' resources and then on their motivations in two subsequent sections. To put these findings in a broader perspective, a comparative section follows. In the final section, we will discuss how our empirical results fit the expectations formulated in chapter 2 and will suggest some refinements of our model.

Data Sources and Methodological Approach

This chapter uses the same data as discussed in chapter 3, but instead of carrying out a country-level analysis, we will zoom in on the party level, using both candidate and party data to systematically compare how and when the parties adopted social media. It may seem strange to use candidate data in party comparisons, but they are important because candidates' use of social media reflects the party culture and strategy, and particularly on Twitter it is the politicians that people follow, not the parties (see chapter 3). A party may well devise a strategy to wage a social-media campaign, but if that party does not succeed in getting a substantial amount of candidates on Twitter, it will most likely be in vain.[1]

In chapter 3, we identified the three bigger parties: the conservative liberals (VVD), the social democrats (PvdA), and the Christian democrats (CDA) (see also Andeweg & Irwin, 2005:47). The three major smaller parties are the progressive liberals (D66), the radical right populists (PVV), and the socialists (SP). The five other parties are minor: two small orthodox Christian parties (CU, SGP), the environmentalist parties (GL, PvdD), and the party for the elderly (50Plus). Compared to the three bigger parties, each of these eight parties is smaller, which is why we will call these eight parties "smaller."[2] The Dutch party system hosts no less than three postmaterialist parties: the progressive liberal D66 and the green parties GroenLinks (GL) and the Partij voor de Dieren (PvdD) (see also Lijphart, 1999:86–87). It also hosts two populist parties: the populist radical right (PVV) and the left populist socialist party (SP) (cf. Schumacher & Rooduijn, 2013; Van Kessel, 2015).

Quantitative data is available on each of these parties' social-media presence from 2010 onward. Specifically we will use the data on all 493 candidates of the 2010 elections, all 512 candidates of the 2012 elections, the 150 MPs in 2015, as well as data on different indicators of professionalism, interactivity, and the use of social media's opportunities. Though we collected most of these data ourselves, points where we use other data are indicated in the text. Evidently, all these different data cannot always be easily compared to each other: some indicators are measured in percentages, while others are measured on scales ranging from 0 to 3.5. Yet the parties' *ranking* on each of the indicators can be compared and that is exactly what we will do in this chapter. In addition to these more quantitative indicators, we will also use our expert interviews to examine which motivations underlie the choice to use social media. They will also provide us information about the available resources. We have held thirteen in-depth expert interviews, talking to eight social-media managers, two politicians who were very active on Twitter and Facebook, one social-media campaigner, one consultant, and one general campaign leader (and national politician). These interviewees represent nine parties (see Appendix 1). The two parties not represented are the notoriously media- (and academia-) skeptic populist party (PVV) and the party for the elderly (though we did speak to an assistant of one of the latter's MPs informally).

Descriptive Analysis: Mapping the Diffusion of Social Media

Typically, the literature makes a distinction between smaller and bigger parties,[3] with normalization scholars expecting that bigger parties dominate the online environment just like they dominate the offline environment and equalization scholars contending that smaller parties are stronger online than offline. Chapter 2 outlined our expectation that in the first phase of diffusion some smaller parties are actually ahead of the bigger ones, after which the bigger parties catch up in the second phase. Other parties keep lagging behind and enter social media only later in the game.

Quantitative Diffusion

Table 4.1 provides information on the parties' presence on social media. It also includes their network size and when they started their accounts on Twitter and Facebook. We also added data for Google+ to gauge diffusion for a "newer" social-media platform as well. Table 4.2 adds information on how many of each party's candidates were present on different social media in 2010 (early adoption) and 2012 (widespread use). Figure 4.1 shows Twitter data on MPs in 2010 (early adoption), 2012 (widespread), and 2015 (when the laggards entered). These data mainly focus on social media's advertisement opportunity.

Based on these three sets of data, three clusters of parties can be discerned: (1) the bigger parties, who catch up with the frontrunners in 2012; (2) a subset of the smaller parties, namely those who are either already active in the first phase of diffusion—the innovators—or had caught up by 2012; and (3) the remainder of the smaller parties who lag behind until, or even well into, the laggard phase of diffusion (2015). Below we will discuss the development for each of these groups in more detail.

Table 4.1 Parties'p resenceo ns ocialm edia

Party	Twitter				Facebook			Google+	
	Account since	*Public/ protected*	*Followers[a] (April 2015)*	*Estimated real followers[c]*	*Account since[b]*	*Friends/likes[a,b] (August 2012)*	*Friends/likes[a] (April 2015)*	*Account (April, 2015)*	*Followers[a] (April, 2015)*
VVD	Jul-07	public	70	27	2010	37	49	Yes	22
PvdA	Jul-07	public	59	19	2010	10	23	Yes	23
CDA	Mar-09	public	19	13	2010	2	11	Yes	20
D66	Apr-09	public	59	36	2009	14	33	Yes	22
PVV	Mar-08	private	2	n.a.	none	n.a.	n.a.	No	n.a.
SP	Nov-08	public	21	14	2009	13	23	Yes	21
GL	Jan-08	public	27	19	2009	17	23	Yes	20
CU	Dec-08	public	11	9	2012	1	10	Yes	0.2
SGP	May-10	public	5	3	2012	2	6	No	n.a.
PvdD	Feb-09	public	12	9	2010*	3	40	Yes	0.4
50Plus	Dec-10	public	0.5	0.4	2012	0.5	3	No	n.a.

[a] Int housands.
[b] Before the general elections in respectively 2010 or 2012, otherwise indicated with an asterisk; Smith (2012).
[c] Estimated by Twitteraudit.com on April 22, 2015.
Source: Manually collected from social media platforms; Smith (2012).

Table 4.2 Percentage of candidates per party who have a social-media account

Party	VVD	PvdA	CDA	D66	PVV	SP	GL	CU	SGP	PvdD	50Plus[a]
Twitter 2010	44	49	36	46	4	22	80	22	7	24	n.a.
Twitter 2012	84	87	96	86	55	67	95	82	33	72	30
Hyves 2010	49	50	67	44	25	42	47	50	7	59	n.a.
Facebook 2012	67	87	81	70	71	80	86	68	40	68	50

[a] 50Plus did not participate in the 2010 elections.
Source: Ownd atac ollection.

Bigger Parties

While the greens were ahead in the early days of social media (2010), the three main Dutch parties, PvdA, VVD, and CDA, caught up with the frontrunners by the time Twitter and Facebook were taking off in the Netherlands (2012). The two biggest of them, VVD and PvdA, were actually the first to sign up on Twitter. Both parties also rank in the top four of the biggest party networks on social media, even if we correct for fake followers (Table 4.1). The VVD especially has

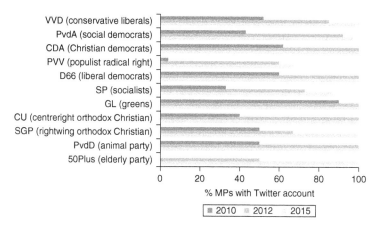

Figure 4.1 Percentage of MPs who have a Twitter account.

a very high number of Twitter followers and Facebook likes—though the Twitter number is less impressive when fake accounts are excluded. The PvdA numbers are impressive when we look at their candidates' Facebook use. By 2012, they had become the number one party in that respect (87%), and the party was also among the large group of most active parties on Twitter (Figure 4.1; Table 4.2). Looking at the table, the Christian democrats (CDA) seem to be somewhat behind, but that would be mainly based on their network size.[4] In terms of using social media, it was actually one of the most visible parties on Hyves in 2010 (Table 4.2). Even though it was relatively late to arrive on Twitter, the party made the biggest jump of all parties between 2010 and 2012 in terms of the percentage of candidates using Twitter. Moreover, it has been in the top three on all platforms ever since (Figure 4.1; Table 4.2). In sum, in terms of adopting social media, the bigger parties were only really behind the tech-savvy, postmaterialist greens in 2010. The PvdA and VVD caught up by 2012 and took over by 2015, while the CDA jumped on the social media train between 2010 and 2012. The major parties might not have been the most enthusiastic from the start, but from the widespread diffusion phase onward, they were among the parties that dominated social media.

(Pro)active Smaller Parties
One Dutch party has been active on social media from the beginning: the Green Party. It outperformed the major parties in the early-adoption phase and remained in the lead throughout the entire diffusion process.

The two other postmaterialist parties, D66 and PvdD, also do particularly well. There is one surprise in the group of smaller parties though: the Christian Christenunie (CU) has done really well from 2012 onward.

In 2010, when social media were quite new, the greens were well ahead of the other parties. Eighty percent of their candidates were already on Twitter, for instance, as opposed to less than 50 percent of every other party (Table 4.3). In 2012 and 2015 they are still in the lead, though they now share this position with other parties (Figure 4.1). In terms of followers and likes, the party also does really well, especially when taking their small vote share into account (see chapter 3). The biggest of the postmaterialist parties, D66, was relatively late to start a party Twitter account (Table 4.1), but in terms of the presence of individual politicians they were in the middle in 2010 and in the lead from 2012 onward (Figure 4.1). The party does especially well on Twitter, but is also quite visible on Facebook. The tiny party for animal rights, PvdD, was somewhat behind in 2010, signing up for Facebook after the elections and with relatively few candidates on Twitter, but by 2012, their (two) MPs were all on Twitter and Facebook, and their percentage of candidates on Twitter and Facebook was almost as high as that of the leading parties. In 2015, their party accounts have very respectable networks on Facebook and Twitter. In fact, their Facebook network is the second largest of all the parties.

So far we encountered few surprises. In line with our theoretical expectations, the postmaterialist and bigger parties did well. However, one of the smaller parties doing well on social media, the CU, is clearly not part of the postmaterialist "usual suspects": It is a traditional center-right religious party. In the early-adopter phase, it was a laggard, opening its Twitter account a year later than the leaders personal accounts launch, only starting to use Facebook in 2012, and showing low numbers of candidates and MPs on Twitter in 2010. By early 2012 however, the party suddenly caught up with the frontrunners (Figure 4.1). Especially the increase in the percentage of candidates on Twitter is noteworthy, jumping from 22 percent (2010) to no less than 82 percent (2012), which puts them on par with the frontrunners. In terms of network size, they are actually rather close to their larger Christian sibling CDA (Table 4.1), even though the latter has about three times as many voters (chapter 3).

Inactive Smaller Parties
Four parties are less prone to using social media. Of these four, the party for the elderly (50Plus) and the ultraorthodox Christians (SGP)

are behind on all fronts, and last to open Facebook and Twitter party accounts. By 2015, both parties are on Facebook and Twitter, but their networks are both minimal, their party accounts are hardly ever used (50Plus), and they do not use newer platforms such as Google+ (SGP, 50Plus). In short, both parties are inactive on social media.

The two populist parties constitute a somewhat more complex cluster of cases. They were early (PVV on Twitter) or have considerable networks (SP) but are clearly behind when it comes to their candidates or even MPs being present on the established platforms; even in 2015, when 95 percent of all the Dutch MPs are on Twitter. These two parties seem the most hesitant and do not seem to stimulate their politicians to use social media. Perhaps this should come as no surprise, as overall— and not unlike their populist brothers and sisters (Mudde, 2007)—they have a hierarchic and centralized party structure, with their strategy revolving around the party leader. Emile Roemer (SP) had the most followers on Facebook in 2012. Similarly, Geert Wilders' personal Twitter account has the highest number of followers from 2012 onward (415,000 followers in 2015) (see chapter 3), but the PVV as a party does not even have a Facebook account. Such selective use of social media makes it nearly impossible to reap the benefits of its unique opportunities. Establishing personal contact with citizens, interacting via multiple channels, targeting specific groups, and allowing individual politicians to become issue owners in the eyes of the traditional media are all nearly impossible this way. Of the two, the PVV is the most centralized and authoritarian in its leadership, and its reluctance to use social media is also more pronounced than the SP's.

Qualitative Diffusion

Adopting social media is one thing, but making use of all its opportunities is quite another. In this section we will discuss to what extent parties make use of social media's four opportunities.[5] Based on the VIRAL data, our expert interviews, manual coding, and external sources, we composed Table 4.3, which provides a series of indicators for professional social-media usage per party. The dotted lines between the rows indicate the three clusters we identified in the previous section: the bigger parties, the (pro-)active smaller parties, and the inactive parties.

Bigger Parties
As reflected by our overall professionalism ranking, the three bigger parties perform quite well in terms of the quality of their social-media

Table 4.3 Indicators of professional social-media usage per party

Party	Overall Professionalism Rank[k]	Professional layout[a]			Advertisement: Activity				Human contact: Outreach & Interactivity			Salon debate	Target group
		Party: Twitter 2015 (0–3.5)	Professional profile picture MPs 2014 (0–3)	Party: Facebook 2015 (0–3.5)	Tweets per month (party 2015)[b]	Tweets per day (all candidates, 2010)[c]	Tweets per day (all candidates, 2012)[d]	Facebook posts (Party, 2012)[g]	% of citizens contacted via social media if contacted by this party (2012)[f]	PTA (people talking about the party) (2012, in thousands)[g]	% of tweets including @-mentions (candidates, 2014)[h]	Mentioned by journalists as posting tweets with the highest news value (2014)[i]	Did parties show a target-group strategy in 2012's campaign and onwards?[j]
VVD	4.3 (3)	3.5	1.8	3.5	76	142	499	17	11%	20	52%	7%	+/–
PvdA	5.0 (4)	2	1.3	2	112	76	516	27	11%	6	59%	10%	+/–
CDA	6.2 (7)	3	1.2	3	137	95	359	41	5%	1	42%	5%	+/–
D66	2.7 (1)	3.5	1.5	3.5	93	77	430	30	11%	6	50%	19%	+
GL	2.9 (2)	3	1.5	3	130	92	441	56	12%	3	64%	10%	+
PvdD	5.8 (6)	3	2.0	3	132	22	99	145	n.a.[f]	1	52%	5%	+/–
CU	5.6 (5)	2	1.6	3.5	73	64	375	134	10%	1	55%	0%	+
SP	7.0 (8)	1.5	1.4	3	64	30	618[e]	26	8%	2	63%	5%	–

				No account	No account			No account	n.a.[f]	No account			
PVV	8.2 (10)	1	0.8			0	37		n.a.[f]		17%	14%	–
SGP	7.7 (9)	2	3.0	3	1.5	3	13	67	n.a.[f]	1	70%	0%	–
50Plus	8.4 (11)	1.5	2.0	1.5	4	n.a.[c]	137[e]	15	n.a.[f]	1	33%	0%	–

[a] Scores based on profile picture and background picture: +1 for logo as profile picture; +0.5 for slogan next to logo in profile picture; +0.5 for background adjusted to party colors; +1 for background logo; +2 for background with volunteers, politicians, or pictures related to ideology.

[b] From opening account until April 9, 2015.

[c] Total for all candidates, not an average, from April 27 until June 8; it reflects the potential visibility of parties on social media; 50Plus has a n.a. value, as the party did not exist at the time.

[d] See (3); from August 11 till September 11.

[e] Averages are disproportionally driven by a few candidates who were so active after the elections that we could not retrospectively count the number of tweets during the campaign. Based on the overall data, these missing data points were replaced by a count of 4,000. This was done for 11 candidates; 5 from the SP (11% of SP candidates); 1 from 50Plus; without this candidate the daily number would be 8 for this party. *Source:* NKO weighted (wgt1c), $n = 1,677$.

[f] Based on NKO 2012; for some parties the group size was too small (<20) to calculate reliable figures.

[g] Based on likes, shares, and comments measured for the last 30 days of the campaign; calculations based on data from Smith (2012).

[h] Based on data from Weber Shandwick (2014:15). No specific time period was indicated other than 2014.

[i] *Source:* Weber Shandwick (2014:15); Based on a survey among 84 journalists from a pool of 200. 26 percent said that none of the parties stood out.

[j] Scores are based on expert interviews. We asked the respondents about the target group they wanted to reach through social media. If they answered journalists, the party received a (–); if they indicated that they wanted to reach a specific group of people but could not back that up with examples of how they did so in practice, *or* also mentioned that they also wanted to reach out broadly, they received a (+/–); if they could identify target groups and present examples, they received a (+). For the PVV we had no interview information, hence we looked at the accounts of the party and the politicians. Based on the following arguments we scored the party a (–): its top Twitter politician, Geert Wilders primarily targets journalists; the party account has virtually no tweets; and the party has no Facebook account.

[k] For each indicator we gave every party a rank score: the best performing party ranking 1, then down to 11. If parties were tied, the rank points were divided (e.g., a tied first position between two parties leads to 1.5 for both). We calculated the averages of all rank scores per main category (professional layout, advertisement, etc.) and subsequently took the average of these five scores. This score basically summarizes the average position and shows the quality of social-media usage relative to the other parties; the scores have no absolute meaning.

Source: Manually collected from social media platforms; NKO 2012; Smith (2012); Weber Shandwick (2014).

use: their party accounts look fairly professional, their candidates send out a reasonable number of tweets, they use the interactive features of Twitter, and a reasonable number of journalists finds their tweets to have the highest news value. The only weak spot for all three seems to be their lack of a clear-cut target-group strategy, but then again this should not come as a surprise, given that the three of them are broad-center or catch-all parties.

(Pro-)active Smaller Parties

Similar observations can be made for the group of (pro)active smaller parties. The two biggest postmaterialist parties are ahead of the pack and are ranked highest on professionalism. Both parties appeal to journalists, have an explicit target group strategy, interact with their followers, and have professional accounts. The progressive liberals (D66) even institutionalized Twitter interactivity, as the party organizes weekly "question time" sessions (D66, MP, 2013). The greens perform surprisingly well among journalists. Indeed, their 10 percent score is well above or equal to the three major parties, and the social-media manager was very conscious about the opportunities social media offer here:

> I do not read any newspapers, but I do follow some journalists. I mean, you can directly contact them on Twitter and get into a debate with them. (GL, social media manager, 2013)

The postmaterialist Party for the Animals (PvdD) performs relatively well too. It is very active on Facebook, has professional accounts, and interacts frequently. The smaller CU also pops up in this list of parties that perform well. While journalists may not find the party interesting, the party has a clear target-group strategy aimed at middle-aged Christians, uses Twitter and Facebook a lot, and has accounts that look good.

Inactive Smaller Parties

The third group of parties includes the same laggards as in the quantitative diffusion analysis: the two very small traditional parties and the two populist ones. The populist radical right PVV basically only uses its party leader's Twitter account as a press release center; the Twitter accounts of the rest of the candidates and politicians have an amateurish look and overall the quality of the social-media use is low. The same is true for the party for the elderly (50Plus). Even though one of its candidates tweeted a lot during the campaign, interaction is rare, and journalists do not believe the tweets are newsworthy. Moreover, the party account pages do not look professional and are rarely used. The

Orthodox-Christian party SGP does slightly better, at least the party and MPs' accounts look professional. The candidates also interact a lot, but the tweets were not perceived as newsworthy, while the party mainly targets journalists via Twitter.[6] Lastly, the populist SP performs badly, though not as badly as the three other parties in this cluster. The party has a few politicians who are extremely active on Twitter, but it does not manage to reach a lot of citizens, has no target-group strategy, and has rather amateurish Twitter accounts.

Explanatory Analysis: Resources

In the previous section, we analyzed how the Dutch parties use social media and discovered that they could be divided in three groups. How can the differences between these three groups be explained? Here we address this question by looking at the resources of each of the parties. Theoretically, the role of resources is not entirely clear, and also in practice we heard that they are both cheap and that money is important. The online communication manager of the CU illustrates the first position:

> Social media are a very simple way, a very cheap way to reach a lot of people. (. . .) we also use advertisements on Facebook. Spending as little as 30 or 40 euro you can already reach several thousands of people. (CU, social media manager, 2014)

The manager of the social democrats puts it as follows:

> [D]uring the campaign we reached more than one million viewers per day through Facebook. That is about as much as one television ad. But the latter is far more expensive. (PvdA, social media manager, 2013)

Twitter is even cheaper to use, as it had few ways to boost a tweet's reach that would give richer or bigger parties an advantage. However, it turns out the importance of money should not be underestimated. Money can still buy you a competitive advantage in terms of buying you a following, buying you a professional social-media team, and boosting the quality of your content. As the social-media manager of the 2012 greens' campaign said:

> A lot of people say "social media are cheap." But if you want to do it well, it does cost money and personnel. (. . .) If you want to make everything look good, it costs money. (GL, social media manager, 2013)

Table 4.4 Resourcesp erp arty

	Money (budget)[a]	Personnel capacity (fte)	Equipment (advanced software)	Journalists (number of contacts)
Major parties				
VVD	High	0.8	Yes	High
PvdA	Medium	1.0	Yes	High
CDA	High	1.6	Yes	High
Active other parties				
D66	High	1.0	Yes	High
GL	Medium	1.0[b]	Yes	Low
PvdD	Low	0.75	No	Low
CU	Medium	0.2–0.3	No	Medium
Less active other parties				
SP	Medium	0.2–0.3	No	Medium
PVV	n.a.	n.a.	n.a.	n.a.
SGP	None	0.4	No	Low
50Plus	n.a.	0.0[c]	No	Low

[a] The budget data are relative to the parties' other communication tools.
[b] The social-media manager was laid off after the 2012 electoral defeat. The online manager is now responsible for social media (+/- 0.2–0.3 fte).
[c] Our informal contact hinted that social media was more of an ad hoc responsibility, loosely coordinated by the party's overall communication manager.

Note: We ranked the parties according to the findings of the descriptive analyses.
Source: Self-collectedi nterviewd ata.

In what follows we examine this tension between cheap social media and the power of money and resources in greater detail. Table 4.4 presents an overview of the resources the parties allocated to social media.

Bigger Parties
The three major parties are clearly allocating money and personnel to social media during and after the 2012 campaign. Only the fourth party in electoral size, the postmaterialist D66, does so to the same extent. We also asked our respondents what they could do with their budgets. Among others, money allowed the VVD to *boost their network size*. On Facebook likes they said:

> *[We were] not buying them in Asia. That would only be cosmetic. We used sponsored stories, whereby photos of people's friends pop up on their timelines saying that this person liked our party's page. Especially after a while, that really worked exponentially.* (VVD, social media manager, 2013)

The party decided to invest in Facebook at a later stage and used these sponsored stories as a fairly expensive way to increase their number of friends or likes from 2,000 to more than 20,000. A "like" costs approximately 1 euro (1.24$) (SP, social media manager, 2013), so the party probably spent more than four times the complete online-campaigning budget of the smaller CU.

Money also helped to buy the bigger parties expertise. The bigger parties have one or more full time social-media managers, supported by technical online staff. Most of the smaller parties, however, only had part-timers. The benefits of having people working on social media full time are clear to the social-media managers. The social democratic party's social-media manager highlights that time allows you to "monitor the other parties," "make sure that everybody gets an answer to their question within the hour," "train candidates," "be available for questions from politicians," and "experiment (. . .) and visit a lot of social-media training events" (PvdA, social media manager, 2013). When a party wants to boost traditional media coverage, monitoring tools and staff also matter. The social-media manager of the social democrats explains:

> *Twitter is like a weather forecast. When you see a journalist tweeting about a certain topic or you see he picks it up from someone, well, then you know the item will possibly be in the news. (. . .) in such cases you need to make sure you have your defense ready. (. . .) The good thing about Twitter is that you are no longer surprised.* (PvdA, social media manager, 2013)

In addition, using more *expensive software*, the party monitored how long their politicians took to answer questions so that they could, when candidates took too long, answer the questions instead. The online campaign manager of the conservative liberals echoes the importance of full-time personnel: "We are constantly reading the news and thinking of things and ways to post it on social media" (VVD, social media manager, 2013). Similarly, the Christian democrats' social-media manager is:

> *Using Tweetdeck, for instance, to filter [messages on Twitter]. I always use it and have search terms to filter all the messages of my politicians. . . . also the bigger [paid] programs, like Coosto, are very valuable [to us].* (CDA, social media manager EP, 2014)

This advantage is further reinforced by the availability of technical support staff who can make infographics or who can Photoshop pictures quickly and professionally. Even if they do not have the staff, richer parties can afford to hire high-level social-media *consultants*. The PvdA did so (PvdA, social media consultant, 2014), and the Christian democrats even got training sessions by coaches from Twitter and Facebook themselves (CDA, social media manager, 2014).

(Pro-)active Smaller Parties

Of the four (pro-)active smaller parties, three have a substantially smaller electorate than the bigger parties, the Greens, the PvdD, and the CU. As a consequence, they do not have the same financial resources as the bigger parties, which means that fewer resources can be allocated to social media (see Table 4.4). Only the greens had a fulltime social-media manager, but he was laid off after the 2012 campaign. The progressive liberal D66 Party had more resources to spend and also allocated them disproportionally to social media:

> *[D]uring the campaign we started to monitor [social media], using software, a [paid] webcare team, and several [paid] external consultants. (...) the campaign manager said "Okay, I want this," and suddenly a lot more money became available.* (D66, social media webcare, 2014)

However, here the modern and tech-savvy postmaterialist cadre also seems to have been crucial:

> *We started to meet as staff and people who were involved as volunteers, [the ones] who had [social-media] expertise.* (D66, social media webcare, 2014)

D66 thus not only used its financial resources, but also used the existing expertise in the party and among its voters and sympathizers. As such, they were able to create one of the first large webcare teams in the Netherlands. Even drawing the interest of television who broadcasted an item the social media team, they could successfully stimulate their politicians to use social media, hire an external consultant, and have an internal team of up to 15 *paid* staffers working on social media. Interns could make professional Youtube videos; the list puller was a devoted Twitter user; professional software was available; the party paid to promote posts; it spread fundraising clips to ask for small donations via text messages; its top candidates were trained in using social media; and a social-media manual was made for the rest of the candidates (D66, social media webcare, 2014; D66, high profile politician, 2013). This unique combination of enthusiasm, available innovative expertise, and money translated into the party catching up in terms of social media by 2012.[7]

In addition to the free expertise of volunteers, politicians, and the cadre of the party, this enthusiasm also explains the extraordinary performance of the other two postmaterialist parties, who had even tighter budgets. For the tech-savvy greens this was particularly crucial in the early-adoption phase. As the greens' social-media manager explains: "[W]e always had to fight for our budget" (GL, social media manager, 2013). But the 2010 list puller Femke Halsema firmly believed in the power of social media and used Twitter to communicate with other candidates on the list (GL, social media manager, 2013). The access to expertise and tech-savvy enthusiastic volunteers helped them to the top early on despite their limited financial resources. The party is very popular among students, and late in the campaign this paid off:

> [T]he last few months of the campaign [in 2012] we had six interns. At that time we had the personnel capacity to push things. (GL, social media manager, 2013)

The regular technical support staff knew their way around Photoshop and could make infographics quickly and professionally, which normally costs more time and money.

Regarding the Party for the Animals, a similar pattern can be observed. Despite having no advanced software, social-media budget, or connections to journalists, they did use their limited funds to pay social-media staff (Table 4.4) and could build on their volunteers as well,[8] particularly an IT specialist who helped expand the party's social-media profile in his spare time (PvdD, social media manager, 2014).

The last party in this group is the small CU. In Section 4.3, the party stood out by catching up with the postmaterialist and bigger parties against all odds. Of the three Christian parties, CU appeals most to Christian young adults, but it would be a stretch to say that expertise is broadly available among their volunteers or politicians. Here, interviewing their social-media manager turned out to be crucial in explaining why they managed to perform so well despite their limited resources (cf. Table 4.4). The party's social-media manager, who has an MA in "new media," started as a volunteer in the 2012 campaign. During this campaign he developed several online communication activities and did some social-media work for the party. Toward the end of the campaign, the party recognized that social media might offer them new opportunities. He showed them that 40 Euros ($50) could boost a Facebook message to 16,000 people. This result was so enticing that the party started to invest in social media and gave him a part-time job during the final stages of the campaign. Later on this turned into a full-time position when a new, younger head of communications was appointed (CU, social media manager, 2014). Budget wise, the social-media manager managed to scrape together more resources every year, resulting in steadily improving social-media use:

> *In the beginning I had to roam the free internet. Now we have 500 euro [or: 615$], and they say, go to Getty images and buy professional photos.* (CU, social media manager, 2014)

However, the budget was limited, and certainly no match for that of the bigger parties. Additionally, he was the only in-house expert. However, he managed to compensate by having the active support of the online team of the EO (7.0 fte), the Evangelical public broadcasting service that is specifically targeted at CU's electorate.

> *We had to learn how to use Facebook. We experimented a lot. (…) I regularly talk to the EO. (…) They do have [access to advanced software]. This generates all kinds of new statistics and a lot more data. That is really useful when experimenting (…) I learned an awful lot from them.* (CU, social media manager, 2014)

In other words, the fact that CU's extraordinary position is truly exceptional is mainly due to one very enthusiastic and convincing staff member who managed to optimally use his contacts in- and outside the party.[9]

Inactive Smaller Parties
The two small parties with traditional electorates (SGP, 50Plus) hardly had any resources (Table 4.4). The social-media manager of the SGP had no budget at all, though he did get "a new camera to take nice pictures" as a consolation prize (SGP, social media manager, 2014). The party for the elderly did not have a full-time social-media expert, and social media was the responsibility of the MP's assistants.

The position of two populist parties again is somewhat harder to interpret. The socialist party spends a moderate amount of money on social media, but this should not be overstated. As the party's social-media manager puts it:

> *We simply cannot afford a lot of the social-media analytical tools. (…) And while we are having this conversation I am missing hours of Twitter feed. (…) We do not have a person covering social media full time. It is a side*

task and perhaps that is a weakness of us, but on the other hand, I have a million things to do. (SP, social media manager, 2013)

Whereas the social-media manager at least took the time to talk to us, the populist radical right PVV simply closes its ranks and does not give any information in this respect. Their social-media strategy is clear, however: it is not used as a party communication tool and only the party leader uses Twitter to provide sound bites. Clearly this is a cheap strategy, one that does not require a great deal of money, but it also does not optimize the use of social media and only works by the grace of journalists willing to play along.[10]

Explanatory Analysis: Motivations

While the results from the analyses of the quantitative and qualitative diffusion process and the role of resources mostly confirm our expectations, some questions remain unanswered. For instance, why did the bigger parties not invest in social media sooner? How important was the need to look modern for the postmaterialist parties? Why are the populist parties so hesitant? Each of these questions refers to the parties' motivations to take to social media. These motivations are the last major building block of our theoretical model.

Bigger Parties
Although the bigger parties find social media highly important nowadays, they were not the first to fully embrace them, and this was no coincidence. As the social-media manager of the conservative liberals states:

> *We are no trendsetters in the sense that we will not help a social-media network grow. We are no early adopters and that is a deliberate choice. We go where our voters are. Once a network starts to take off, we'll go there. But we will not invest time and energy in it when our voters are not there.* (VVD, social media manager, 2013)

And he deliberately compared the VVD to the greens:

> *"[for] GroenLinks [the greens] it fits how they want to be perceived: being hip and modern. We are modern, but not (...) in the way that the "medium is the message."* (VVD, social media manager, 2013)

Similarly, the social democrats (PvdA) also arrived relatively late: Only by August 2011 did they hire a full-time social-media expert. Before, the party did not want to invest too much as it did not believe in social media, only hiring somebody to work on social media "on a very small contract" (PvdA, social media manager, 2013). The social-media manager of the Christian democrats echoes this: "We had to convince them and show the value of it (...) it took a while" (CDA, social media manager EP, 2014). In sum, the three major parties only really started to use, and invest in, social media once it was clear that they were useful and were used by a substantial part of the electorate. Given the resources these parties have, such a strategy is sensible, as they could simply buy their way in at that point.

(Pro-)active Smaller Parties
In the previous section, we showed the importance of informal expertise and enthusiasm for the postmaterialist parties. The enthusiasm shown by party leaders Alexander Pechtold (D66) and Femke Halsema (GL) well before the 2010 elections, as well as the enthusiasm of large groups of interns and volunteers both already indicate that the motivation to take to social media was present from the start. Both parties cultivate an image of being modern and creative, and indicate that social media fit such an image. The greens even saw social-media use as "the celebration of being modern" (GL, social media manager, 2013) and the online D66 campaigner we spoke to said that the party

> *saw the big companies starting to use social media and founding webcare teams. It was taking shape and we thought: "Okay, this is what we want from now on." It is certainly a way to showcase yourself. (...) you can create visibility, become a trending topic or something like that. I think it is a bit symbolic [of being modern] indeed."* (D66, social media webcare, 2014)

The Party for the Animals (PvdD) was also very aware of the possibilities social media offered to reach their electorate directly. The party's spin doctor and webmasters were enthusiastic about social media, and already in 2010 they hired a social-media manager, because they wanted "someone who was always online, who simply likes it and who is devoted to it" (Van Esch, 2014). The party also feels that communicating through social media is a very natural means of communication, as "a lot of [their] voters are active on social media. It also has to do with the fact that a lot of animal welfare news is shared on social media" (PvdD, social media manager, 2014).

The situation of the small CU is clearly different: It did not strongly desire to be modern. The motivated volunteer was simply in the right place at the right time, as the party had also just hired a new head of the communication department, who "thought social media were very important. And this changed our internal organization immensely. (. . .) It was a genuine mind shift" (CU, social media manager, 2014). Additionally, at that time Facebook especially started to also be used by middle-aged Christians, the core electorate of the party. All in all, this does suggest that motivated individuals can make a difference if they are not hindered by structural barriers (cf. opportunity structures).

Inactive Smaller Parties
The motivation of the two smaller inactive parties was quite straightforward. The ultraorthodox Christian SGP simply used social media, especially Twitter, because some politicians "noticed that this was a way to get into the newspapers," end of story (SGP, social media manager, 2014), while 50Plus hardly uses social media at all.

We did not get access to the PVVs motivations because they refused to be interviewed, but their exceptional strategy of not being on social media as a party and primarily using the Twitter account of their leader suggests that for the PVV the main advantage of social media is found in attracting attention from the traditional media (see Table 4.4 on their success in this respect), but that it does not want to stimulate individual candidates to use social media (see next chapter). This fits their overall centralized party organization and strong leadership culture.

To a lesser extent, a similar vibe was present in the interviews we held with the SP campaign leader and their social-media manager. On the one hand, the party itself believes that social media fit its identity well: "It fits the SP, it fits our way of working. We work on a small scale, focused on people themselves" (SP, social media manager, 2013). And the campaign leader also stated that: "It is our duty to be a creative party!" Yet he was less convinced of the importance of social media:

> *I think people overestimate social media. (. . .) Television debates are the most important (. . .) It's all about the good idea and good ideas pop up more often when you bring young people together with a slice of pizza than when you invest in an expensive team of experts.* (SP, campaign leader/senator, 2013)

At the same time, the social-media manager wanted to be creative, but he was too overburdened to fully make use of social media's possibilities: "I have 100,000 other things to do." And he was also highly skeptical about journalists and about reaching the SP electorate via social media:

> *The Twitter community of journalists and politicians, it's all incredibly in-crowd (. . .) Our electorate is not active on Twitter during [television debates]. (. . .) But it shows how within the journalist community everybody is locked up in their own world. (. . .) They are all active on Twitter all the time, and then assume [Twitter] tells you something about public opinion.*
> (SP, social media manager, 2013)

Moreover, the party was not just skeptical about social media's importance and its opportunities, the party was also very conscious of its potential *negative* effects. It did not want to stimulate its politicians to use social media because the party wanted to keep control over its politicians and the main message:

> *It happens. We have had MPs who go off the deep end. Paul Ulenbelt [highly active MP on Twitter] has done that several times, which makes you think, "Boy oh boy, is that necessary?" You know. And then sometimes we call or e-mail. Like, "Hey, was that really a smart thing to do?"* (SP, social media manager, 2013)

Later on he added:

> *[W]e do not want PvdA-like incidents, where individual MPs all air their own individual opinions and start criticizing each other on social media.*
> (SP, social media manager, 2013)

More than other parties, populist parties seem to consider social media as a potential threat to their party discipline and organization.

Comparative Perspectives

Above we have found that in the Netherlands postmaterialist parties make good use of social media and outperform non-postmaterialist smaller parties, while the bigger parties can catch up quite easily because of their resources. But how do these findings compare to those in other countries? One of the exploratory research questions that we formulated in chapters 2 was: *How does social media's impact on the power balance depend on the political context?* To provide a tentative answer to this question, we compare our conclusions about the Dutch case to those of other studies.

The Netherlands Compared to Other Western Democracies

While most studies on Web 1.0 argued for normalization (for a summary see: Gibson & McAllister, 2014), some studies argued for equalization in the case of social media (Chen & Smith, 2010; Gibson & McAllister, 2011; Lassen & Brown, 2011; Southern, 2015). However, these studies were often based on a very small number of smaller parties. We have shown that for some small parties social media (Web 2.0) might invoke equalization (postmaterialist parties), while bringing normalization to others (non-postmaterialist and especially populist parties). Taking a closer look at these other studies shows that their results are actually in line with our findings, but that their conclusions perhaps claimed too much, as the model presented in this book can explain their results just as well. For instance, Vergeer and Hermans stating that the hypothesis that "less established and smaller parties [in the Netherlands] will use Twitter more extensively, is unsupported" (2013:414) are correct but this conclusion is misleading at the same time. Indeed, half of the smaller parties used social media less than the bigger ones, but the other half used it as often and even just as professionally. Their conclusion might be read as a sign of normalization, but it should not necessarily be.

Similarly, a recent study on the Austrian 2013 elections concludes that "[c]andidates from both the extreme left and right are less likely to make use of online campaigning [including Twitter and Facebook] (. . .) In the Austrian context the reticence of populist radical-right candidates is especially eye-catching given that FPÖ leader Heinz-Christian Strache is the "Internet star" among the country's politicians" (Dolezal, 2015:115). On the one hand, this is a strong indication that our results on populist parties and the tension between the social-media presence of their party leader versus the rest of the party are no exception. On the other hand, our analyses and model help to reinterpret this result of Dolezal's study as a consequence of the hierarchical nature of these parties, not of their extremism.

Gibson and McAllister's seminal 2014 study also uncovers results in line with our findings. They find equalization at first for the one smaller party in Australia—the Greens. This result fits our reformulation of their multistage model, making their expectations conditional on the type of smaller party. In particular they find that minor-party candidates' "use of digital campaign tools is linked to an improved electoral performance and thereby they support the hypothesis that cheaper social media technologies are helping minor parties to become more competitive" (Gibson & McAllister, 2014:10, 13). Given that the study includes one, postmaterialist, smaller party only, this finding fits our model quite well.

Moving on to the United Kingdom, we see that in 2013 67 percent of Labor MPs used Twitter, 56 percent of Conservative MPs did so, while this percentage was 79 percent among the MPs of the (postmaterialist) Liberal Democrats (Martin, 2013). Regarding the 2010 general election, Southern (2015) shows that in highly competitive districts the adoption of Twitter and Facebook was equally high among the Liberal Democratic, Green, and Labor candidates (after controlling for budgetary resources), followed closely by the Conservatives. The populist BNP and UKIP were lagging behind, which is in line with our analysis' results as well.

American Exceptionalism?

In many countries with a majoritarian electoral system, the third party is not strong enough to compete. Such parties may be so marginal that they cannot compensate for their lack of resources with some expertise from their electorate, volunteers, or cadre. If this pool is simply too small, even modest equalization appears unlikely.[11] The American case is very illustrative in this respect. The American general elections are generally a battle between the Republicans and Democrats, whether they involve Congress, the Senate, or the White House itself. In this context, social media's impact will be weaker, and thus the enthusiasm among the third parties is likely to be lower. Indeed Evans, Cordova, and Sipole (2014) find that third-party candidates in the United States were less present on Twitter. In other words, the normalization taking place in the United States is rather exceptional and should not be generalized to proportional systems or majoritarian systems with a higher number of effective parties such as the United Kingdom.

Does this mean that social media did not transform power relations in American politics? Not necessarily. As we have demonstrated in this chapter, focusing only on bigger versus smaller parties leads to a highly distorted view. Power relationships among bigger and among smaller parties can also change due to the introduction of social media. In the American context, this was reflected in the victory of the Obama campaign, which made clever use of social media. It did see social media as part of the old media logic, but utilized them in a way that fit social media best: creating networks and mobilizing specific groups (Katz, Barris, & Jain, 2013; Parmelee & Bichard, 2013). This should not be mistaken for equalization. As extensively discussed in this chapter, professional use requires technical tools and lots of staff, and thus is far from cheap.

In the Netherlands, the (cheap) salon-debate function is more dominant. To understand these different foci, it again is important to consider the context. In the Dutch context, smaller parties target traditional media because it is an important way to get into newspapers and on television; buying loads of broadcasting time is beyond their budget and broadcasting time is partly available on major public channels. In the American media and political system, reaching an audience through television is mostly done by ads, news stations are more strongly aligned politically, and there are big campaign budgets. More generally, the budgets are disproportionally large due to liberal campaign-financing legislation (allowing for more staff) and, particularly during presidential campaigns, interns and volunteers are applying from everywhere, so there is an abundance of personnel capacity.

The takeaway message here is that in the fairly atypical American system (Almond et al., 2003:777) resources can be expected to be far more important in shaping power relations among parties. In such a context, social media will only have a limited impact on the divide between smaller and bigger parties. Surely, to fully support this claim, more studies of American candidates' motivation for (not) using social media are crucial, but there are few theoretical reasons to expect differences.

All in all, our model seems applicable to many Western democracies: proportional systems and majoritarian systems that have more than two parties. Based on our comparison here, however, we might expect that in the latter—already more personalized (Suiter, 2015)—systems the salon-debate opportunity is less dominant in the parties' media strategies, and the human-contact and target-group opportunity may be more important.

Conclusion

This chapter—a case study of the diffusion of social media and its effect on the interparty power inequalities—presented the first systematic test of our motivation-resource-based diffusion model. Below we summarize our results and compare them to the expectations formulated in chapter 2.

While some authors have claimed that "the two hypotheses—innovation [or equalization] and normalization—are mutually exclusive" (Vergeer, Hermans, & Sams, 2011), it seems that empirically this is not the case. Equalization and normalization are both part of a diffusion process, wherein the *relative* power relations between the bigger and smaller parties are shifting differently depending on the diffusion phase. Moreover, equalization and normalization are dyadic concepts, comparing one party to another: the adoption of social media by one party might lead to equalization versus some parties but, at the same time, might lead to normalization vis-à-vis others. That is exactly what we found. Based on all the material and interviews we collected, we have got a pretty strong grasp of the underlying mechanisms as well, and our motivation-resource-based diffusion model so far is largely confirmed.

We found that in the early-adoption phase equalization took place for the postmaterialist parties and particularly for the greens. This party was ahead early on, as it had the motivation to appear modern and the enthusiasm to do so because its cadre, volunteers, and politicians were and still are tech savvy. Smaller parties such as these might not have many financial resources, but in the early-adoption phase money does not seem the most important factor, as the basic functions of social media are indeed cheap. In terms of staff, they had the expertise and enthusiasm among their ranks and did not need to hire new people. The other smaller parties did not have the motivation to start using social media in this phase of diffusion, and the bigger parties were playing the waiting game, as their electorate was not really active on social media. In sum, Expectation 2 is confirmed: In the early-adoption phase of social-media diffusion, equalization took place for the postmaterialist parties vis-à-vis the rest. However, Expectation 1 is not fully confirmed, as not all of the smaller parties take up social media even though the bigger parties did lag behind.

In the phase of widespread diffusion, the dynamics change and resources play a much bigger role. We found that the bigger parties started to adopt social media because their electorates were on social media now, and because they had proved to offer important opportunities, an important one being covered or quoted in old media. The bigger parties were capable of catching up with the postmaterialist parties because they had the resources to "buy" expertise, networks, support, and quality content. This confirms Expectation 3 and hints at normalization compared to the remaining smaller parties. Those parties were not strongly motivated to use social media to its full extent, either because it might lead to a loss of control for the central party leadership (populist parties, see Mudde, 2007:270) or because their electorate was not on social media yet (the smaller traditional parties). Regarding the latter, even if they wanted to become more active on social media, they lacked the resources, which confirms Expectation 4. There was an exception, however: If a smaller party is able to tap into external resources and that party's gatekeepers "accidently" are social-media fans, they can catch up and do relatively well, as illustrated by the small center-right CU. Yet this was clearly a deviant case.

The Netherlands moved into the last phase of political social-media diffusion in 2014–2015. At that time the laggards also—somewhat reluctantly—adopted social-media strategies, mainly as channels to increase their coverage in traditional media. But the lack of resources and/or intrinsic motivation still prevented them from really catching up, as formulated in Expectation 5.

In the long run, social media seem to be to the advantage of parties that can tap into resources and that have an identity fitting these modern media (postmaterialists), but it will certainly not shift the power balance drastically, as financial resources help control the damage to the bigger parties. Overall, our results are robust across the analyses: those of quantitative diffusion, of qualitative diffusion, and of the resources and motivation all exhibit the same patterns. Moreover, it is particularly telling that the parties who share certain characteristics show similar behavior: both populist parties are hesitant; all three postmaterialist parties do well; all three bigger parties only start devoting resources to social media in the second phase. In other words, the Dutch case provided an excellent test of our framework and helped to disentangle certain mechanisms.

On a somewhat more general note, the interviews revealed that several social-media managers have a great deal of knowledge about the nuts and bolts of social media, but they also indicated that resources are often lacking to really ensure the quality of social-media use. Other campaign managers were still in the process of exploring and experimenting. This is also reflected in our analyses of how social media are used. While the advertisement and salon-debate opportunities are widely recognized by all the parties, significant progress can be made in terms of using social media to interact with citizens, to target specific groups, and to create long-term relationships. Above all, traditional media are the core focus of many (if not all) parties. In that respect, social media are considered to be a way to provide journalists with information, quotes, and name recognition, on the one hand, and on the other, they are used as a barometer or "echo chamber" (CU, social media manager, 2014) of what will be in the news tomorrow so that parties have their reactions ready. In short, social media are still not used to their fullest potential.

CHAPTER 5

Intraparty Relations: David versus Nabal and Abigail

Introduction

Social media provide individual politicians with tools to build an electorate on a personal basis, which "might lead to less control over the politician . . . Party discipline . . . could subsequently become compromised" (Vergeer, Hermans, & Sams, 2013:481). Politicians who seek their own electorate based on nongeographic and personal politics might benefit from this in particular (cf. Karlsen, 2011). Twitter offers a cheap opportunity to build a large personal public of citizens, NGOs, and journalists that individual politicians can inform about their political actions and ideas. Facebook enables politicians to build a personal network and to share and discuss personal experiences, events, and grievances, but if used that way it is a labor-intensive medium. Some ordinary—not high-profile—candidates may have more incentive to take to social media than others. Echoing the second-wave feminist rallying slogan "the personal is political," our attention is then drawn to politically underrepresented groups such as women and ethnic minorities. In their analysis of the innovation phase, Vergeer, Hermans, and Sams (2011:488) concluded that social media were not actively used for intraparty competition during the innovation phase. But is this still true?

This chapter focuses on our third research subquestion (chapter 2): How do social media change the power balance of individual politicians within parties vis-à-vis each other and politicians vis-à-vis the party organization? More concretely, we will address whether the potential of social media to personalize politics and the opportunities offered by social media to empower underrepresented groups actually translate to empirical reality. To start, we will provide information about our data and research design. Afterwards, we will focus on personalization and underrepresented groups. In each section, we elaborate on the most important concepts, such as the *concentrated* and *diffuse personalization* of politics, and *political representation* and *intersectionality*, examining their implications. Afterwards we compare our findings with studies on other countries. Finally, we will draw more general conclusions about the way social media equalizes or normalizes intraparty power relations.

Data and Methodological Approach

In this chapter, we disaggregate the data introduced earlier to the level of candidates and politicians, in as far as possible, including the data on presence, network size, and posts in different years, as well as the data on being followed by journalists in 2010. The interviews also included two high-profile "social-media" MPs, and we discussed politicians' independence explicitly with the parties' social-media campaigners. This chapter's analyses are also used to highlight a few select politicians to illustrate social media's general mechanisms, also helping us to generate more fine-grained theoretical mechanisms as part of our theory-generating approach.

Most studies on equalization versus normalization focus on the interparty level and distinguish between bigger and smaller (or even fringe) parties. If we move from parties to politicians, we thus first have to establish what exactly constitutes a bigger, smaller, or fringe politician. In what follows, the clustering of politicians is based on their list position: first the list puller, then the other top-ten candidates, and lastly the rest. Clearly a position in the top ten means something different across parties, depending, for instance, on the number of MPs the party has. Focusing on fewer candidates, however, would make the analyses very sensitive to outliers. In interpreting the results, we did keep this in the back of our minds though, and the interviews allow for a more party-specific definition of top candidates. We also take into account whether politicians were already well-known based on previous positions they might have held,[1] and in-depth case knowledge allows us to take into account political crisis and exogenous shocks.

Regarding representation, we distinguish candidates based on their sex and ethnicity. First, ballots and self-reported sex were leading.[2] Regarding the second, we focus on being part of a picture- and name-based *visible* non-Western ethnic minority. This corresponds to the public discourse and categorization of politicians based on seeing them on television and online, or reading their name in newspapers and on ballots. Concretely, ethnic-minorities candidates mostly have a Turkish, Moroccan, or "Overseas territories" (e.g., Suriname, Dutch Antilles) background reflecting labor migration and Dutch colonial history.

Social Media as a Driver of Political Personalization

Theoretical Perspective: The Implications of Social Media for Political Personalization

The personalization of politics gets quite a bit of attention in both political science and communication sciences (Colomer, 2011; Karvonen, 2010; Kriesi, 2012; McAllister, 2007; Poguntke & Webb, 2005; Van Holsteyn & Andeweg, 2012; Van Aelst, Shaefer, & Stanyer, 2012; Vliegenthart, 2012). At the core of the process of personalization is a shifting focus from institutions and organizations to individuals—individualization—which can go hand in hand with more attention for the private lives of politicians instead of their professional lives—privatization. This process can actually lead to two forms of personalization: concentrated and diffuse personalization. The former echoes mechanisms of normalization while the latter echoes equalization.

1. *Diffuse personalization: equalization.* Social media have unique characteristics (see chapter 2) that require a new style of political communication, one that is "personalised, decentralised and unsupervised" (Vergeer, Hermans, & Sams, 2013:480–481). This particularly fits the idea of diffuse personalization: a shift toward individual candidates. All candidates can use cheap and easy-to-use social media, and lower-ranked candidates who have few other campaigning resources might benefit from using social media in particular. Possible spillover effects via traditional media (the salon-debate opportunity) further strengthen this advantage because the central party no longer has a gatekeeper privilege and because journalists are part of politicians' Twitter or other social-media networks (see chapter 3; also Lasorsa, Lewis, & Holton, 2012; Peterson, 2012:432; Weber Shandwick, 2014). This suggests intraparty equalization, because it benefits the candidates with fewer resources and no prior access to media.

2. *Concentrated personalization: normalization.* There are also reasons to expect more concentrated personalization, however: a shift from organizations or institutions to the parties' faces and political leaders. If parties apply a campaign strategy particularly supporting

prominent politicians, the focal point of personalization will be the already powerful politicians. As a more complex medium, Facebook in particular requires expertise and technical support. Political parties can support politicians by creating slick infographics and by sponsoring posts (i.e., pay to boost the reach of their messages). At the same time, parties can discourage lower-ranked politicians from using social media. Additionally, because parties often want to keep control over the central message (especially during campaigns, Norris et al., 1999), social media might increase the probability of slipups or internal divisions getting out in the open, especially when journalist are expected to be on the lookout for controversial tweets, also from lesser-known politicians (Peterson, 2012). All in all, this would result in concentrated personalization (or presidentialization) if the focus were solely on the political leaders (Jacobs & Spierings, 2015; Van Aelst, Schaefer, & Stanyer, 2012), which is akin to the normalization of intraparty relationships.

Yet when can we expect which type of personalization? One can assume that the way a party is organized strongly influences the overall personalization trend, because a party's loss of control and discipline is particularly problematic for centralized parties (see chapter 4). The complexity of the more personal social-media platforms and the efforts of centralized parties might steer the personalization process to concentrated personalization and thereby normalization. Below we assess these processes of personalization empirically.

The Social-Media Dilemma: Party Leadership versus Individual Politicians

Virtually all parties realize that social media, and particularly Twitter, are tailored to individual users instead of institutions. As a campaigner of the progressive liberals noted: "They tweet with Alexander Pechtold [D66's list puller], not with D66 [the party]" (D66, social media webcare, 2014). The number of followers of the two types of accounts reflects this. The most popular politicians have more connections on Twitter than their party accounts. For instance, the PvdA and VVD list pullers had approximately four times the number of followers on Twitter than their parties (October 2013). On Facebook however, parties often have more followers than politicians, including list pullers (see chapter 3). Politicians found Facebook more troublesome and demanding, as they have to "accept friendship requests" and "include pictures and all these things" (Pvda, MP, 2013). Even when parties tried to convince their politicians to use Facebook, they encounter trouble: "[Facebook] does not talk to them. They underestimate the impact" (SGP, social media manager, 2014), and "Facebook is still used more as a private medium by politicians" (VVD, social media manager, 2013). Personalization is linked mostly to Twitter.

This personalization creates a new dilemma for parties: additional media attention and reaching a broader public *versus* losing control and running the risk of slipups. In the words of the social-media consultant of the social democrats:

> *Of course, this leads to tension between the party line and individual opinions, which has become even more precarious because it is indeed easier, faster, and that makes it difficult to communicate consistently.* (PvdA, social media consultant, 2014)

This draws particular attention to the salon-debate function of Twitter. The conservative liberal social-media manager was clearly looking for a balance:

> *Journalists have their Twitter lists ready, they are waiting (...) for a [politician] to make a mistake. That is very nerve-racking process for us. We must talk [to our politicians] and raise awareness.*[3] (VVD, social media manager, 2013)

The greens' social-media manager simply "surrendered":

> *You just have to let go of a lot of control. You're not the boss anymore. Not the only one who has connections with [traditional] media.* (GL, social media manager, 2013)

How parties perceive this loss of control differs considerably. The parties have three "strategies" to deal with social media: discouraging their politicians, monitoring them, and/or training them. In the first columns of Table 5.1, we categorized parties by their overall degree of centralization and summarized the parties' attitude toward intraparty use of social media. The parties' positions range from mainly conservative and risk averse (three parties) to stimulating social-media activity (five parties). Three are more ambiguous or have no clear policy.

1. *Forbid, discourage or abstain from stimulation.* We find the most strongly centralized parties at the extreme end. One party (leader) strongly discouraged (or banned) its MPs from voicing their views in any way (PVV) and the other populist party (SP) stressed that not

Table 5.1 Centralization, social-media policies, and politicians' usage per party

Party	Centralization	Quote	Party stimulated politicians to open an account	% MPs (candidates) on % MPs (candidates) on Twitter[b]			% Candidates on Facebook
				2010	2012	2015	2012
PVV	Very High	"MPs are not supposed to excel, but to be invisible."[a]	No[a]	4 (2)	60 (55)	67	71
SP	High	"We told our MPs: 'only use Twitter if you feel good confident doing so. (…) Don't do because you supposedly have to.'"	No	33 (22)	73 (67)	83	80
SGP	High	"[It is] a lot of fuss, and may cause internal fights" yet "they can reach a large audience if they use it"	Ambiguous, with different points of view in the party	50 (7)	67 (33)	100	40
50Plus	High	n.a.	n.a.	n.a.	50 (30)	100	50
PVDD	Modest to High	"It is not an obligation in our party (…) I prefer good and active use over signing up and not using the account."	No, but a bit ambiguous due to training	50 (24)	100 (72)	100	68
CU	Modest	"Two MPs are very active [on Twitter]. The rest has an account out of decorum, but hardly use it."	Yes (MPs considered it to be a commitment)	40 (22)	100 (82)	100	68
CDA	Modest	"I had to show it [Twitter] to them, more like, go ahead, start using it."	Yes	62 (36)	100 (96)	100	81

VVD	Modest	"As a party we use it [Twitter] quite scarcely (…). Our politicians and MPs, on the other hand, use it all the more."	Ambiguous, advice is given, but the party also keeps an eye on the MPs	52 (44)	85 (84)	100	67
D66	Modest	"I signed up to Twitter when I became a candidate, partially because the campaign team explicitly requested us to do so. (…) They regularly request us to Tweet more frequently."[c]	Yes	60 (46)	100 (86)	100	70
PvdA	Low	"We do approach people, [but] the [party] is a Twitter-enthusiast party."	Yes	43 (49)	92 (87)	97	87
GL	Very low	"Things do not simply happen. You have to stimulate people to spread news. Some do so on their own initiative; others you need to… I would not say 'force' but stimulate."	Yes	90 (80)	100 (95)	100	86

[a] Based on other sources, particularly PVV MPs who left the PVV delegation and criticized the party for not allowing any democracy and deviating opinions, not even behind closed doors. This quote comes from MP Hernandez, who left the party in 2012. *Source*: Nu.nl (2012) Others also left the party over its undemocratic practices, such as former prominent MP Hero Brinkman. *Source*: Zantingh (2012).

[b] See Figures 4.1 and 4.2.

[c] Ironically, the politician saying this added: "As far as it is possible to give us orders, because MPs are rather obstinate." (D66, politician, 2013).

every MP had to be on Twitter. The SP believed that Twitter was dominated by an in-crowd made up of journalists and the political elite (SP, campaign leader/senator, 2013). The Orthodox Christian SGP also tended to stress the disadvantages more than the advantages. It explicitly legitimized not using social media, at least until the early and late majority phases. By 2015, however, even two of these parties "surrendered" (see 2015 user rates in Table 5.1). At the other extreme, we find parties who actually stimulated their politicians to sign up for and use social media, mainly Twitter. These include all the least centralized parties and several of the modestly centralized ones. For instance, one D66 politician told us that she and other MPs were asked to sign up and that they are monitored to see if they really tweet actively, something that was also confirmed by the party's campaigner.

2. *Monitor and control.* More generally, all parties seem to monitor the Twitter streams of their MPs, particularly during the campaigns.

> *Look, there were two of us. If you have access to good monitoring software that lists all tweets [of the politicians], it becomes a simple matter of staying on top of it. You immediately notice who has not responded [to tweets].* (PvdA, social media manager, 2013)

The social-media managers of centralized parties talk about monitoring more in terms of genuine control and mention speaking to politicians who tweeted something the party did not like (SP, social media manager, 2013—quoted in chapter 3). Among the modestly centralized parties, we found another way of dealing with discord: they informally coordinated who tweets about what and when (VVD, social media manager, 2013; D66, MP, 2013). A tendency to control, however, can spike in all parties. Back in 2012, things did not go well electorally for the least centralized party, as conflicts surfaced during internal elections for a new party leader. MPs tweeted about it, feeding bad publicity. The "information officer, who was completely overworked because of this mess, started to call people all the time saying 'You are not allowed to say that!'," as the social media campaigner told us, adding: "Well, you should never do that, but it was all that stress" (GL, social media manager, 2013). Finally, parties can also control content by actually posting for politicians instead—as is very common in US politics: Hillary Clinton for instance uses the sign "-H" for messages she posts herself, constituting only a small minority of "her" tweets. In four of the parties social-media staff tweet for politicians, and for Facebook this is the case for at least eight of the parties.[4]

3. *Train and support.* Another means of "damage control" or of making sure the party benefits as a whole is training and support, which we found in many parties, but less so among the most centralized ones. D66 had courses for their leading candidates and a guide for local politicians, CDA organized workshop with people from the social-media companies themselves, CU created a list of ten illustrated dos and don'ts, the PvdD's social-media managers trained local list pullers and several other politicians, and so on.

All in all, the parties want to remain in control of the communication flows, but it is clear that social media pose a serious challenge, and the degree to which the parties interfere differs depending on the party's culture, organization, and electoral tides.

Who Uses Social Media: Leaders or Backbenchers?

In this section we focus on the politicians themselves. Figure 5.1 presents the percentages of each cluster of candidates per party on Twitter and Facebook in 2012. By that year, using social media was no longer an exception, but the platforms were not yet politically saturated (see Table 5.1). The closer a party is located to the top right-hand corner in Figure 5.1, the more candidates were on that platform; a position above the diagonal indicates that the top-ten candidates are more present on social media than other types of candidates (more concentrated individualization); a position below the diagonal indicates the reverse (more diffuse personalization).

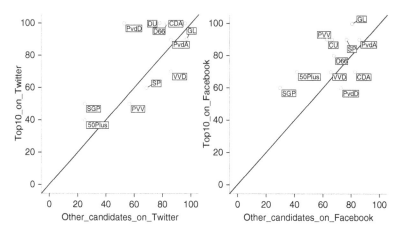

Figure 5.1 List pullers, top-ten, and other candidates on social media in 2012.

All the list pullers have accounts and most numbers indicate concentrated personalization. On Twitter, this is the case mostly for 50Plus, SGP, PVV, and to a lesser extent the SP, which largely overlaps with the discussion above. At the same time, it has to be noted that even among those parties, 25 to 60 percent of the other candidates are present on social media, even if this is purely "out of courtesy" as one interviewee said. For Facebook, none of the parties is found in the lower left-hand quadrant. We can also note that most parties are above the diagonal—16 out of 22 instances—indicating that top-ten candidates are more often on social media than other candidates, with differences ranging from 4 to 47 percentage points on Twitter, and 4 to 30 percentage points on Facebook. Not surprisingly, this gap was found most often for parties that we found actively stimulated MPs using Twitter. In general, the presence of politicians follows a cascade logic: all list pullers, a large part of the top-10 candidates, and a substantial part of the other candidates use social media. The difference is bigger for more centralized parties on Twitter than for less centralized parties. The two figures suggest that the top politicians remain ahead but that many other candidates are also trying to create their own public or network.

Building a Public: Advertisement Opportunity

What about that public? The average number of followers (4,000 in 2012) is highly skewed by a few very popular candidates followed by 50,000 to 250,000 Twitter accounts (see chapter 3). Of the top five 2012 candidates on both Twitter and Facebook, four were list pullers. More generally, Figure 5.2 shows that people follow higher-ranked politicians most, and the list pullers particularly stand out. Moreover, the public follows the list pullers because they are list pullers. For instance, the three candidates who became list puller in 2012 and were already on Twitter in 2010 all doubled or tripled their number of followers, even though they were already top-ten candidates in 2010. The only new list puller who was not on Twitter in 2010 (Van Haersma Buma) gathered 15,000 followers in a short period, which is more than the average top-ten or other candidate—although it has been noted this particular politician "bought" followers (see chapter 3; NOS, 2012). The party leaders further consolidated their lead: the simple registered number of followers having more than doubled again since 2012 (see Table 5.2), and even controlled for fake followers, the number in 2015 is higher than the gross number of followers in 2012.

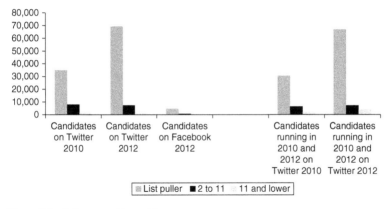

Figure 5.2 Followersp erl istp osition.

Table 5.2 Number of (real) followers of political leaders in 2015

Party	Political leader	Estimated real followers in 2015	Registered followers in 2015
VVD—Conservative Liberals	Mark Rutte[a]	239,000	398,000
PVV—Populist Radical Right	Geert Wilders	210,000	419,000
D66—Liberal Democrats	Alexander Pechtold	159,000	353,000
PvdA—Social Democrats	Diederik Samsom	97,000	159,000
SP—Socialists	Emile Roemer	57,000	108,000
PvdD—Animal Party	Marianne Thieme	31,200	47,800
CDA—Christian Democrats	Sybrand Buma	22,100	33,300
CU—Center-right Orthodox Christian	Arie Slob	21,600	34,800
SGP—Ultraorthodox Christian	Kees van der Staaij	19,300	30,800
50Plus—Party for the Elderly	Henk Krol	8,600	11,900
GL—Greens[b]	Bram van Ojik	8,500	10,500
Average		79,000	146,000

[a] The account he uses is the Prime Minister's twitter account, @MinPres.
[b] Only a few weeks after the data were collection Bram van Ojik stepped down as party leader, and Jesse Klaver became the new leader of the Greens. Jesse Klaver had 22,400 reported followers and 19,600 estimated (real) followers on May 25, 2015.

Source: Self-gathered data from Twitter, May 1, 2015; the real number of followers are estimated using Twitteraudit.

For the other candidates it is much harder to build a following, and it takes more time, as is also illustrated by the relatively small difference between top-ten and the other candidates (Figure 5.2). At the same time, building a substantial following *is* possible, as shown by the outliers who have a larger audience than some (including their own) list pullers. Moreover, the candidates who were on lower list positions in both 2010 and 2012 still managed to double their average follower size to an average of 4,000 (Figure 5.2),[5] which constitutes a larger and more interested public than that they could reach in town halls or by canvassing all day. Clearly it does take time. As the VVD's social-media manager said:

> On Twitter it is quite hard to build a follower base. It is a slow process, unless you are a well-known politician (...) when you are new MP, or merely a candidate, it is a long-term process. (VVD, social media manager, 2013)

Yet merely having a follower base is not enough; to build a connection, politicians need to post content. For Facebook there seems to be an algorithmically optimal number of posts on average of one a day, and it is not about sheer quantity of posts (e.g., CU, social media manager, 2014). On Twitter it is a matter of showing up in people's Twitter feed and thus tweeting very regularly. In this respect, list pullers are again the most active: 22 tweets a person per day in 2012, while the top-ten candidates show an average of 11 and the other candidates 9. However, there are important exceptions, for instance, both Pieter Omtzigt (CDA) and Mei Li Vos (PvdA) being among the (10%) most active candidates in 2012.

Building a Public: Salon-Debate Opportunity

As stressed by the campaign managers and politicians, it is not all about the number of people in one's network; it is also about who these people are. A spillover effect via journalists to traditional media is crucial here. In the words of an MP of the progressive liberal D66:

> I notice that journalists pick up discussions on Twitter really quickly. Then it features in the national news coverage, and it becomes a hype, while this might not have been the case otherwise.

We have data on how many prominent journalists (out of a list of 15) followed each of the candidates in the 2010 election. These data show two important patterns. First of all, the most-followed candidates did include several of the list pullers, but some list pullers were only followed by three or four of the journalists. The list pullers that were followed most were social-media innovators such as Femke Halsema (greens) and Alexander Pechtold (progressive liberals). Second, among the candidates followed by ten or more journalists (14 candidates in total) were many who were known for their social-media activity,[6] including incumbent and fairly well-known MPs placed relatively low on their party's list such as Mei Li Vos (PvdA) and Pieter Omtzigt (CDA). Politicians who were active in the early-adoption phase and had already built a strong media network seem to benefit from that later on (such as the two aforementioned candidates, who both got elected with a high number of preference votes). There were relatively few of such candidates, however. The average top-ten candidate was followed by fewer than two journalists and for all the other candidates this number was roughly 0.5, while list pullers were followed by an average of 5.5 journalists. We have to add though that this reflects the early-adoption situation. By now social media are more widespread and especially Twitter is more integrated in journalists' daily routines. As such, more politicians are likely to be followed by more journalists.

High-quality Use

So far we have focused on the advertisement and salon-debate opportunities, but the human-contact and target-group opportunities can also yield intraparty benefits. Unfortunately, there is only a limited amount of data available to check these opportunities. Below we list the available information.

With the shift to widespread diffusion by 2015, the quality of posts and profiles can be expected to become more important, with party support and training concomitantly becoming more important as well. Comparing the profiles of all 150 MPs of 2014, we find surprisingly few patterns of professionalization. Of the MPs with the most professional photos, only three are list pullers (Thieme, Van der Staaij, and Roemer), and most are unknown MPs; only four list pullers have a background picture (Thieme, Pechtold, Wilders, Krol); and only the CU's list puller stresses his personal life, with his profile reading:

> Husband, father (in-law); CU parliamentary delegation chair; likes music, reading, athletics, PEC Zwolle [first division soccer team], and lots more.

There are also potential advantages to having fewer followers. The aforementioned politicians Vera Bergkamp and Pia Dijkstra (both D66, see chapter 3) tweeted moderately, about seven to eight tweets a day in the 2012 campaign, but above all they really engaged with their public during and outside campaigning periods. Both received quite a lot of preference votes too. Having fewer followers does have the potential benefit of getting more genuinely interested followers and being able to interact with them. If the number of followers is too high, genuine interaction is nearly impossible. As such, having a moderate follower base may be more effective.

The more successful MPs on Twitter and Facebook do seem to have a strategy as to whom they target. They tweet interactively and focus on certain topics. This holds for Bergkamp and Dijkstra, who focused, among other topics, on LGBT issues; Vos, who targeted self-employed professionals; and Omtzigt, who focused on representing the province of Twente. In each of these cases, it seems that social media were part of a larger strategy and resonated with the general political style of the candidate. Not surprisingly, candidates who started out of the blue in politics but were running personal campaigns did not manage to obtain many preference votes (more on that in chapter 7). This indicates the necessity of having a strong network or of being well-known before the campaign starts.

Personalization Yes, But How?

Overall, parties and politicians agree that Twitter fosters individualization. Facebook fits the idea of privatization best, but that is actually one of the reasons why politicians use it less for politics. Parties anticipate this personalization process by either discouraging and monitoring or training and supporting their MPs, whereby it seems that high-profile MPs get most attention. This is also reflected in the data on signing up for social media and building a (media) network or public: list pullers are often leading on social media, far ahead of other top-ten candidates, who are closer to the lower-ranked candidates. This suggests a process of very concentrated personalization (normalization).

However, the support for the process of normalization needs to be nuanced in at least two important ways: (1) the relative progress of low-ranked politicians and (2) the success of several high-profile politicians who use social media professionally. Both these nuances rest on the importance of professional or high-quality use of social media, which has become more and more important as the presence of MPs on social media becomes more rule than exception.

1. *Relative position*. Candidates who started to use Twitter in the early stages of diffusion particularly managed to build a network of several thousands of followers over the years, which is more than they would have reached *without* social media. List pullers and top candidates would have gotten media attention and followers regardless. Relatively speaking, social media thus offer opportunities for the lower-ranked candidates who start early, keep posting messages, gather an audience of some journalists, and, above all, patiently persevere. In other words, some equalization seems to take place, though the degree of this is quite minimal on average.

2. *Successful low-ranked politicians*. Some candidates were relatively high-profile at the time they entered politics, or managed to build a profile first, for example, as alder(wo)man of a major city. For them it is easier to build or extend their network, and if they use the opportunities offered by social media these politicians can actually create a personal power base. In some cases, they even outperform the list pullers in terms of professionalism. Not surprisingly, most of these candidates are found among the postmaterialist parties, though not exclusively so. A few are among the most centralized or traditional parties. These exceptions illustrate that equalization is possible (though rare).

Social Media as an Ungated Arena for the Underrepresented

Representation

One specific form of intraparty competition related to the concept of privatization is the representation of different groups in society.[7] Two core politically underrepresented groups are women and ethnic minorities (e.g., Celis et al., 2008, 2014; Childs & Lovenduski, 2013; Ruedin, 2013; Saward, 2010; Weldon, 2002). Both sociocultural groups are underrepresented in the political system and often roam in the margins of (traditional) parties. As a result, they often receive less support from these parties. One can thus wonder whether social media mirror such offline inequalities (normalization) or compensate for them instead (equalization).

The opportunities created by social media (see chapter 2) make social-media platforms especially interesting for socioculturally marginalized groups to voice their political views and establish linkage with citizens whose vote they seek or whom they claim to represent. First of all, underrepresented groups might seek to compensate for their under-representation (in both politics and the traditional media) by funneling their efforts to the more accessible social media. This might actually lead to a more equal representation in the formal institutions if they manage to build an online power base. The salon-debate opportunity of social media is particularly interesting here, given the absence of the party-level gatekeepers that play a role in the overrepresentation of ethnic-majority men (cf. Blumler & Kavanagh, 1999; Enyedi, 2008). The advertisement opportunity is also interesting, as it has been suggested that social media can contribute to changing stereotypes and narrowing representation gaps (Bailey et al., 2013). Moreover, social media can draw attention to shared experiences based on people's marginalized gender and ethnic identity (Celis & Spierings, 2014). In this respect, the target-group and human-contact opportunities of social media seem to fit politicians belonging to underrepresented groups more than any other societal group.[8]

Social media thus offer unique opportunities to underrepresented identity-based groups in society, but whether these opportunities actually materialize into something real remains to be seen. Below we examine Dutch women and ethnic minorities, doing so in an (intercategorically) intersectional way (McCall, 2005; Spierings, 2012). We argue that the difference between men and women might be different for ethnic-majority candidates than for ethnic-minorities candidates. For instance, women were underrepresented 2 to 1 among Dutch parliamentary candidates in 2010 and 2012, but among ethnic-minorities candidates there was no clear underrepresentation (see chapter 3).

Posting Underrepresentation Away?

Table 5.3 shows the percentages of candidates on social media split by year, sex, and ethnicity. First of all, these figures illustrate that ethnic-majority women are more present than ethnic-majority men, but that ethnic-minorities men are more present than ethnic-minorities women for all years and platforms. Second, Hyves and Facebook seem to favor the politically marginalized groups more than Twitter, which fits Facebook's more private nature as a personal pegboard, whereas Twitter already is more encapsulated in the political dynamics.

Additional logistic regression models however mainly show that the found differences are strongly linked to interparty differences and to the candidates' age. Filtering out the impact of those factors transforms the relative advantage marginalized groups have into underrepresentation. Women and ethnic-minorities candidates are more strongly represented in the social democratic and the green party, and in fact they

Table 5.3 Candidates with a Twitter and Hyves/Facebook account by ethnic status, sex, and election year (%)

		Twitter		Hyves (2010) / Facebook (2012)	
		Man	*Woman*	*Man*	*Woman*
Ethnicity	*Majority*	32.9	38.2	45.2	46.7
		73.3	*79.4*	*73.6*	*80.0*
	Minority	29.4	21.4	52.9	50.0
		81.8	*75.0*	*86.4*	*80.0*

Note: Nonitalicized figures indicate the percentage of candidates on Twitter or Hyves/Facebook for each group in 2010; the *italicized* figuresf or2 012.

Source: VIRAL; own calculations; *n*=1 ,024.

might choose to run for these parties because those parties have a longer history of minorities representation, identity politics, and explicit emancipation agendas. Similarly, ethnic-minorities candidates might decide not to run for the strongly conservative and centralized parties such as SGP and PVV, which leave little room for identity-based diversity and representation given their ideology and organizational structure (see chapter 4; Spierings et al., 2015).

For Twitter we can also make a longitudinal comparison, which shows that the increase in usage was stronger among ethnic-minorities candidates than ethnic-majority candidates, among both women and men. This development might indicate a lack of motivation among ethnic-minorities candidates in the first phase of diffusion, which seems to fit previous results in chapter 4 that suggested their electorate was less present on social media. However, of the eight ethnic-minorities candidates on Twitter in 2010, six had a follower base well above the overall median. Later in the diffusion process, a larger number of people from an ethnic-minorities background were present on social media, and it seems the potential of social media started to kick in. Intriguingly, in the widespread diffusion phase, the politically dominant category of ethnic-majority men (almost 60% of all MPs) was the *least* present of all four groups. This might be attributed to them feeling less of a need to use social media because of their privileged position in the offline world.

The position of ethnic-majority men may even be "worse" if women and ethnic minorities use social media more actively than white men, as suggested by, for instance, Evans, Crodova, and Sipole (2014). Our data do not show such a clear pattern though. Figure 5.3 presents each group's mean and median number of tweets a day during the

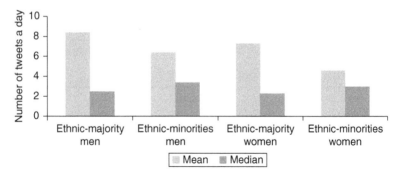

Figure 5.3   Twitter activity of 2010 and 2012 parliamentary candidates.

campaigns.[9] According to the means, ethnic-majority candidates are on average more vocal on Twitter than ethnic-minorities candidates, but the medians strongly suggest that these differences are caused by a small number of very active (12 of 569) ethnic-majority candidates. Among the "less massively popular" candidates, ethnic-minorities politicians are even slightly more active. After accounting for the highly active candidates, men and women tweet about equally often. A look at the median number of followers for 2012 (not shown here), nuances the gender differences even further: it seems the underrepresented sex might be on Twitter more often, whereas the dominant sex attracts more followers. By 2012, the median ethnic-majority candidate also attracts more followers than the median ethnic-minorities candidate.

Finally, in terms of professionalism, little to no secondary data are available on the human-contact and target-group opportunities that can be disaggregated by sex of ethnicity. We can compare the Twitter *profiles* of the 150 MPs at the end of 2014, and if anything we find more professional accounts for ethnic-majority MPs and no gender differences. Also, none of the 15 ethnic-minorities MPs had personal background pictures; neither did many of the ethnic-majority candidates, though men slightly more often than women. All in all, there is little evidence for the application of these more specific social-media opportunities to gender and ethnicity, and we also are rather cautious in drawing conclusions, as the content of messages is not taken into account here. Delving more deeply into widely covered incidents in Dutch politics (see Box 5.1) does show that identities are sometimes linked to salient programmatic issues, which might mutually reinforce each other and lead to a social-media equalization for the politicians involved.

Box 5.1 Ethnicity profiling on social media

An interesting test case of using social media to interact and target niche groups based on ethnicity is provided by politicians that have implicitly made an issue of ethnicity. In this respect, three social democratic MPs of Turkish descent who have been in conflict with their party on the issue of integration provide some intriguing bits of evidence: Tunahan Kuzu, Selçuk Öztürk, and Keklik Yücel. The first two were forced to leave the party and the last stepped down as spokesperson on the topic of integration.

After she stepped down, Yücel tweeted little about integration, mostly sending out messages about gender equality and LGBT emancipation, her new dossiers. On Facebook these issues dominate as well. Kuzu and Öztürk, however, founded a new party "Denk NL"—In English: Think Netherlands—and tweet and post on Facebook about ethnicized issues, including integration, but they do not strongly link this to their own experiences. Kuzu uses a more personal style and provides political background information on his Facebook, but more as an advertisement opportunity. Indeed, despite the many comments and shares, Kuzu does not interact with his followers very regularly.

The Twitter networks of these three candidates are average at best: 4,327, 2,573, and 3,758 followers, respectively (May 6, 2015). These numbers are around the 2012 median, but to 2015 standards they are quite low. The news coverage of their critique on the party or them leaving the social democrats seems to have boosted their network considerably. On Facebook, Yücel has a moderate 1,400 followers, Öztürk 5,000, and Kuzu (chair of the 2 headed delegation) 16,000 (May 6, 2015), which constitutes a noteworthy network.

Kuzu and Öztürk both have a migration background and clearly build their social-media presence around the issue of integration, and quite successfully it seems; however, the human-contact opportunity is not used strongly.

Sources: Klompenhouwer, 2014 and NU.nl, 2014.

Equalization for Some, Normalization for Most?

When it comes to whether the opportunities that social media offer are used in ways that advance politically underrepresented or marginalized groups, the results seem to be mixed. In the first as well as the later phases of diffusion there are some indications of equalization: on the one hand, women use social media more, as do ethnic-minorities candidates. Moreover, women do address gender-equality issues, but this could also be due to the fact they are mostly candidates for parties that fight for gender equality. On the other hand, and feeding normalization, women neither tweet more, nor do they have more (journalist) followers. The women who address gender equality do so "politically"; we did not come across women who clearly link it to their own experiences. In terms of motivation and resources, we can tentatively deduce from our material that two things are crucial: (1) there appears to be little motivation to really connect with the public "as a woman," and (2) the main resources are (a) being a list puller, who are generally men (17 of 21 studied), reproducing the unequal gender power balance, or (b) people's own expertise and tech-savviness. Both "resources" imply that party and ideological subculture are important.

The pattern is different for ethnic inequalities. At first, ethnic-minorities candidates were clearly less present on social media and only had small followings containing fewer journalists (normalization). They did not have big incentives to be active on social media, probably because their electorate was largely absent from social media in the early-adoption phase, and they did not have a lot of party resources. However, when social media became more widespread, the number of followers and activity of ethnic-minorities candidates increased disproportionally. In that phase, they did particularly well compared to other non-high-profile candidates who are part of the ethnic majority (equalization). Also, we found clear indications that the politicization and saliency of migration and integration politics made ethnic-minorities candidates take quite visible positions on these ethnicity-related issues. Though there are no strong clues that they consciously strategize using the human-contact opportunity of social media, they are building relatively large communities on Facebook, and the integration debate can help them to make a push through "old" media to become better known and recruit a larger following. In other words, social media do seem part of a larger (media) dynamic.

Comparative Perspectives

The Netherlands Compared to Other Proportional Systems in Europe

We can draw empirical material and conclusions on intraparty equalization and normalization from three studies on other European countries: Jacobs et al.'s (2015) study of Belgium, Dolezal's study on Austria (2015), and Karlsen's study on Norway (2011)—all three were carried out in times of the early-adoption or at best early-majority phase. Though the bits and pieces in these studies that relate to our study of the Netherlands provide provisional support for the external validity of our model to proportional systems, these studies do not allow us to test whether our model also holds once the social-media diffusion process has advanced.

First of all, our concentrated personalization conclusion resonates with these studies. Each of them shows that the public focuses on list pullers in particular, and the latter are more present on social media early on. The Belgian study, for instance, shows that list pullers have up to 30 times as many followers (Jacobs et al., 2015:17). The studies also indicate that parties remain important—Karlsen particularly stresses this for Norway. Digging a bit deeper, the results of these studies support, or can be explained in terms of, our additional "equalization for some" conclusion: some professional, postmaterialist, and well-known politicians do stand out. The Austrian and Norwegian studies include party affiliation and show that the relatively young candidates from postmaterialist and social-democratic parties[10] have the highest likelihoods of using social media in the early phase of diffusion. In both cases the centralized PRR and communist as well as traditional (agrarian, Christian, and thus older) candidates are less likely to embrace social media.

From the Belgian and Norwegian studies we can derive that motivation alone is not sufficient to build a strong network, and only a small group of politicians will stand out from the masses. In Belgium, only a very limited number of candidates uses social media professionally (in this case interactively) (Jacobs et al., 2015; see also Jungherr, 2014), and in Norway only a small part of the candidates (11%) was more candidate focused than party focused in their communication (Karlsen, 2011). Jacobs et al. (2015) also suggest the importance of traditional media attention for building a following: the list pullers gained the most followers in the campaign, most of all the least well-known list pullers it seems; in other words, the candidates that were flagged as important by the parties and media, but yet not broadly known by the public. All in all, these pieces of information overlap with our conclusion that social media will mainly benefit a select group of more professional and motivated candidates who know how to exploit the opportunities offered by social media.

One major difference between the three countries discussed above and the Dutch case concerns the relative importance of the different social-media platforms. For Norway, Karlsen (2011) finds that candidates use social media most as an advertisement tool, and slightly less so for its salon-debate opportunity, which fits the fact that they are more politically active on Facebook than on Twitter. The same holds for Austria: More politicians signed up for Facebook than for Twitter (Dolezal, 2015). This raises the question whether there are differences in the media systems that might explain this. Answering this question is beyond the scope of this book, but our analyses suggest that it is crucial to examine whether journalists in these countries are also more on Facebook than on any other platform and whether the candidates focus more on target-group and human-contact opportunities. Moreover, the role of party support and campaign-finance regulation should definitely also be considered in this respect.

Privatization, Gender, and Majoritarian System Such as the United States

Above we focused on party-list European systems. The literature on the United States and other majoritarian systems has also paid little attention to intraparty competition, just like the literature on proportional systems. This is not surprising, as preference voting plays virtually no role in majoritarian systems and often there is only one candidate per party running in a district.[11] Still, other cleavages might play a role in electoral bodies, such as gender and ethnic inequalities. This section will focus on these two topics.

For the Dutch proportional system, we found that the differences in candidates' social-media use by gender and ethnicity were small and largely due to party-related factors, and also in other countries women were not found to be clearly more active (Dolezal, 2015; Larsson & Kalsnes, 2014; Luhiste & Sudulich, 2015). In this respect, we should remember that in the Netherlands social media were hardly ever used in a privatized way, which fits the country's political culture. In countries like the United States, politics is already much more privatized; partners, children, and grandchildren are generally part of the campaign, for instance. In the Netherlands, a single 48-year-old man is prime minister, which seems nearly unimaginable in the United States.[12] In majoritarian systems, social-media use probably leans more toward human-contact strategies. Given the more male-dominated nature of majoritarian system, this can be expected to be particularly advantageous for underrepresented groups who can appeal to the experiences they share with large parts of the electorate—an accepted political practice in the United States. Unfortunately, several of the studies on the Twitter adoption of US members of Congress do not distinguish between men and women (Lassen & Brown, 2011; Peterson, 2012) or there were no female candidates running (e.g., the republican presidential primaries [Conway, Kenski, & Wang, 2013]). The only study that focuses on the Twitter use of US House candidates (Evans, Cordoba, & Sipole, 2014) shows that women are more often active on Twitter and, if so, also more often use it for political means, though not more personally or interactively. A study on the UK looking at gender differences (Southern, 2015) also shows that women are more likely to have a personal website, but it only covers a subsample of highly competitive, and thus unrepresentative, constituencies for Twitter and Facebook. All in all, however, the evidence on majoritarian systems points toward women being more prone to start using social media.

Conclusion

Most studies examining the diffusion and effect of social media in politics adopt an implicit intraparty perspective, analyzing individual politicians' adoption and usage rates (and at best aggregating these data to the party level). The role of parties is often left out of the picture— probably due to data limitations. Yet typically such studies also apply an interparty theory to interpret their findings: the equalization versus normalization framework. There is thus a tension between *intraparty empirics* and *interparty theories*. In this chapter we suggested two solutions. The first way to integrate the intraparty and interparty dimensions is by adding a theory on the (changing) relationship between parties and politicians to the mix: the personalization theory (Van Aelst, Shaefer, & Stanyer, 2012). The second involves applying the logic of the equalization-normalization debate to what is happening at the intraparty level. This in turn opens a range of new questions, such as: who benefits most from social media, top candidates or the more disadvantaged ones? Candidates from traditionally overrepresented sociocultural groups or the more disadvantaged ones? In short, this chapter focused on our third research subquestion (cf. chapter 2).

Personalization: Three Strategies to Deal with the Danger of Decentralization

In their 2011 study, Vergeer, Hermans, and Sams (2011:488) suggest that Twitter as a decentralized medium "might lead to less control over the politician...and party discipline...could subsequently become compromised." Indeed, personalization processes through which the attention shifts from parties to politicians may well be strengthened by the increasing popularity of social media. Both parties and politicians acknowledge that personalization is taking place and is being facilitated by social media. But while the focus shifts from parties to individual politicians, being the list puller of a party remains the main way to draw and keep people's attention on social media (cf. concentrated personalization), and it is largely the parties that determine who the list pullers are.[13] Nevertheless parties also realize that social media offer opportunities but can also challenge their party communication. To deal with this danger of decentralization resulting in cacophony they have established different strategies to deal with this. We discovered three such strategies: (1) discouraging politicians from using social media, (2) monitoring and controlling them, and (3) training and supporting them. This way, parties try to relieve the tension between the party and the politician. The more centralized a party was, the more it interfered with its politicians.

No Surprises: Equalization for Some, Normalization for Most

Within the parties, some politicians are clearly better at using social media than others. In the early diffusion phase, it was mainly the professional, postmaterialist politicians who were active on social media along with the parties' list pullers, but the former were typically ahead of the latter. Politicians of more traditional and centralized parties were found less on social media, as were ethnic-minorities candidates and men. This fits the two core motivational aspects defined in the previous chapters: (a) politicians who are tech savvy and simply like new communication tools (even if these technologies still have to prove themselves) use social media first and (b) the other politicians simply go where their electorate goes (e.g., traditional, older, and ethnic-minorities candidates [see Quintelier & Theocharis, 2012; Schlozman, Verba, & Brady, 2012; Valenzuela, Park, & Kee, 2009]). Of the list pullers, those of the postmaterialist parties often stand out and were found to be social-media enthusiasts; the others are generally online because they need to be and are followed by many people and journalists simply because they are list pullers.[14]

In the later diffusion phases, a lot more politicians are present on social media, which is in line with the idea that politicians follow their electorate. Yet now presence no longer suffices and the quality of social-media use becomes more important. A new list puller can easily catch up in this phase and become part of the leading group on social media, partly because they typically receive support from the party in the form of expertise, attracting extra followers, training, or even extra hands to post tweets for them. Moreover, such candidates are by definition interesting for journalists, which increases their media exposure and social-media attention. Other politicians entering the fray in this phase have to work harder to attract attention, though those who receive media coverage and grab this opportunity to hold on to their new followers fare better. The new resource here, one could say, is having a high degree of political-social capital. Candidates who do not receive a lot of media attention are in a tight spot, as building a network on social media takes time: It is a slow process. The candidates who already were on social media in the early diffusion phase do relatively well, as they have gained expertise and have been building their network for some years.

All in all this implies equalization for some but normalization for most. Indeed, some candidates can now attract more media attention through social media and are able to spread their message more widely, something that was previously impossible in the offline world. In that sense, the playing field is leveled for them.[15] However, we should not forget that this group of candidates is fairly small.

Underrepresented Groups: Progress, but Baby Steps at Best

Our analyses do show some very modest progress: more women than men have social-media accounts, yet they post as much as their male counterparts but have fewer followers. Ethnic-minorities candidates were also typically found on social media more often, had fewer followers, but posted (way) more often. A small case study on the three most visible ethnically Turkish MPs in the Netherlands showed the importance of other media. Crisis and conflict can function as an exogenous window of opportunity. This factor punctures the equilibrium and propels the politician to the forefront in traditional media. Such a dynamic is most likely to be found in topics that are politically salient such as integration and LGBT issues (see Box 5.2), but not gender equality. Internal crises in a party on such issues can draw journalists (though it might damage the politicians' position within their party). Whether such differences between ethnic-majority men and the other groups also level the playing field in electoral terms is mainly dependent on how important the role of followers is vis-à-vis the role of posting. This is something we delve deeper into in chapter 7.

Box 5.2 LGBT identities as target group and human-contact exception

One apparent exception to the rule that politicians are hesitant to make full use of the human-contact opportunity can be found in examples across this book on a social group that we did not focus on here (because it is less underrepresented in Dutch politics, though it also concerns identity issues): gay (f/m) people and politicians. Several politicians openly and strongly identify as gay, such as Boris van der Ham (D66), Vera Bergkamp (D66), Jeroen van Wijngaarden (VVD), pledging their allegiance to LGBT emancipation on their Twitter accounts by, for instance, including rainbow signs (Tanja Jadnanansing, PvdA). These politicians often come from the liberal or postmaterialist (wings of) parties, and large parts of their electorate are tech savvy and present on social media early on. Additionally, several LGBT issues have been hotly debated in politics and on social media since the 1990s, including rather personal and private issues for LGBT people: marriage, adoption, religious schools firing gay teachers, physical sex change or corrections, and legal sex change. This might explain the strong presence of these politicians on social media as well as their relatively large followings. This strong (international) online network of LGBT activists and politically interested citizens might also explain why both our examples on interactivity in chapter 3 were from politicians who were gay or spokespeople on LBGT issues, as they can quickly and easily acquire valuable information through social media. Their network extends beyond being a mere advertisement tool; it is a means of acquiring high-quality information.

Future Research: Some Gaps to Fill

The conclusions above are sometimes still sketchy, as not all pieces of the puzzle were present and reality is quite messy. Yet the mechanisms discussed above do represent the general picture and help us to pinpoint the most crucial elements that shape candidates' resources and motivations, and consequently determine their position on social media. Future studies can build on this.

Specifically, our analyses suggest that a target-group strategy is at the core of success, and the human-contact opportunity is little used but potentially successful if (well-known) politicians genuinely interact with their followers. Content analyses and interviews with politicians can shed more light on this.

The complex relation between traditional and social media deserves more attention as well. Does a candidate's network size really expand after she gets media attention? What leads journalists to follow lower-ranked politicians, and which politicians manage to really retain their new followers? More research is also needed on how politicians use the opportunities offered by Facebook in particular. Do politicians on Facebook focus on advertising themselves or do they try to build a strong personal community?

Finally, we have mainly focused on gender in this chapter, but some other groups do seem to benefit from the diffusion of social media, such as LGBTs and some ethnic-minority candidates given the saliency of integration politics. As such, we propose that future research applies our general mechanisms to the particular political context in a country to see which groups potentially and actually benefit from social media (e.g., for which issues are politically salient, see Schlozman, Verba, & Brady, 2012).

CHAPTER 6

Social Media Go "Glocal": The Local and European Arenas

Introduction

In line with the lion's share of studies in the equalization-normalization debate (Gibson & McAllister, 2014:2; Larsson, 2013; Larsson & Svensson, 2014), the two previous chapters analyzed the consequences of social media on the main or "first-order" political arena (national politics) (Reif & Schmitt, 1980:8). In this chapter, we study whether the conclusions from chapters 4 and 5 also hold in sub- and supranational elections—the "second-order" political arenas. This serves two purposes. First, this helps to establish the robustness of our results. Second, there are good reasons to expect that the adoption and impact of social media is similar but of different strength in second-order arenas. After all, voters might feel there is "less at stake," which results in lower levels of political participation and better chances for smaller parties. Though second-order arenas are influenced by the national-level dynamics, they have their own dynamics as well. Specifically, different coalitions may be in charge or other players may be competing (Reif & Schmitt, 1980:9–10). Second-order arenas typically draw less attention from traditional media, which increases the added value of social media in terms of informing the public (advertisement opportunity) and mobilizing specific electorates (target-group opportunity). Let us take a closer look at local and European elections to see whether this really is the case.

Some Expectations: The Local Level

In the local arena, fewer resources (lower budgets and less professional staff) might particularly advantage postmaterialist parties and postpone normalization, dynamics we identified in chapter 4. Additionally, local elections cover a smaller geographical area, which makes face-to-face and direct communication a viable alternative to media coverage. This in turn may well postpone the adoption of social media, except for intrinsically innovative parties (cf. Larsson, 2013). In short, there are fewer incentives for non-postmaterialist parties to use social media. Lastly, as Reif and Schmitt (1980:11) note, second-order arenas sometimes involve different players while at the same time displaying an important degree of overlap, which creates "interconnections." Indeed independent local parties do exist.[1] These parties can be expected to be relatively disadvantaged, as they, unlike local branches of national parties, cannot tap into the expertise, knowledge, and funding of a mother party. We thus expect a slow diffusion process at the local level, with an even slower normalization process and a stronger equalization effect for the postmaterialist parties.

Some Expectations: The European Level

Some of the local-arena dynamics we discussed are also at play in the European arena. European politics also yield less attention from traditional media (De Vreese, 2001). However, unlike their local brethren, European politicians cannot compensate by taking to the streets and communicating with their constituents directly. Moreover, European politicians typically have more funding and support, while at the same time there are fewer of them, which makes their PR efforts more manageable. Compared to the local level, the motivation of European politicians to use social media is higher (as they have fewer alternatives) and resources are more abundant. This would suggest the early-adoption equalization effect for some to be extra strong. At the same time, the limited attention from traditional media can lead to more incentives for all the European politicians, suggesting a tendency toward normalization. We can thus expect a quick diffusion process resulting in normalization (relatively early on).

Data and Methodological Approach

Though the analyses in this chapter follow the same logic as the previous ones, there are some differences on the data end, because less data is available for the local and European level, although the VIRAL project does include some unique bits of data covering European candidates and local Dutch political actors.

Local Politics Data

On March 19, 2014, local elections were held in almost all municipalities of the Netherlands. As no data are available on the 2010 elections, the analyses here are synchronic comparisons of the 2014 elections. The data we use are derived from two subprojects of our larger VIRAL project (the first covering 4 municipalities and the other 2).[2] For all six municipalities we have data on all the candidates of parties that won seats in 2014 (1,540 local candidates in total). The data on the number of posts refer to the campaign period: the first subproject covering 20 days before the election, the second covering the 7 days leading up to the election.[3] Data is also available on the parties' presence on Twitter and Facebook. The subproject focusing on the four municipalities included in-depth interviews and content analyses of parties' Facebook accounts. The five interviews with local campaigners (see Appendix 1) were modeled on the template we used in chapter 4. Our analysis of the Facebook accounts covered both the quality of the information and the degree of interactivity. This coding is based on the posts themselves as well as the level of engagement with them, following the methodology of Gibson and Ward (2000) and Small (2008). More information about the coding is provided in Table 6.3 presenting the results.

The municipalities are: Amsterdam, Groningen, Houten, and Wageningen; and Almere and Nijmegen. The Dutch capital (Amsterdam) is by far the largest city in the Netherlands (approximately 800,000 inhabitants). Almere, Groningen, and Nijmegen are three other top-ten cities, with 170,000 to 200,000 inhabitants. Houten and Wageningen are considerably smaller municipalities (40,000–50,000), but still they are among the 33 percent of municipalities with the highest number of inhabitants. The municipalities were selected to ensure their similarity on several core demographical levels. They have relatively high-educated populations (37% to 48% has a higher education degree), the turnout in the 2010 elections was between 53 percent and 59 percent, and the percentage of elderly people (>65) only varied between 11 percent and 15 percent.

The exception is Almere, with only 26 percent of its inhabitants being higher educated and only 9 percent of its population over 65. Almere is included because it is one of only two municipalities in which a local branch of Geert Wilders' PVV participated. Regarding the political characteristics of the municipalities, each saw at least one local party win seats, and all three major parties as well as the two main postmaterialist parties participated in all elections. Amsterdam, Groningen,

Nijmegen, and Wageningen are university cities, and particularly D66 and GroenLinks have a relatively strong presence in the councils, while CDA is weaker. Houten is located in the Dutch Bible Belt, with a relatively strong presence of the two small orthodox Christian parties: taking up 4 of the 29 seats, with another 4 going to the Christian democrats.

All six cases can thus to some extent be considered "most likely" to find widespread social-media use. If we find social-media adoption and use already lagging in these municipalities, it is likely to be even worse in the other ones. This case-selection strategy has two distinct advantages. First, it ensures that there is something to analyze: We are likely to find social-media use and managers in municipalities such as these. Second, because the local branches of one and the same parties take different power positions relative to the national situation, these analyses are very suitable to see whether ideology, cadre, and culture are really that important, as we claimed in chapter 4, or whether such factors are merely a proxy for the (local) size of the party, as most of the literature would contend.

European-politics Data

To assess the European arena we can make diachronic comparisons, as we are able to draw from previous studies on social-media use in Dutch EP politics (Vergeer, Hermans, & Sams, 2011, 2013), comparing their data to information about the 26 MEPs after the 2015 EP elections, as well as the European party accounts (if any) of the parties in the EP. Because the number of MEPs is relatively low, it was possible to do an explorative screening of the social-media content as well. Moreover, one of the social-media managers we interviewed (cf. chapter 4 and Appendix 1) was active at the European level, and we explicitly discussed the relationship between national and European politics with her.

Local Politics

In this section we will first discuss the pace of the overall social-media diffusion process at the local level, after which we dig deeper into the inter- and interparty effects, using the national analyses as points of reference to compare with our local findings.

Subnational versus National Diffusion

In the 2014 local elections, Twitter adoption among candidates ranged from 30 percent to 44 percent in the five smaller and medium-sized municipalities, which is about half of the adoption rate in the 2012 (!) national elections. In Amsterdam, more candidates were on Twitter, but the capital's percentage still was well behind the 2012 national level: 62 percent versus 76 percent. Similarly, the average and median number of tweets in the five smaller cities most closely resembles 2010's level of activity for national candidates, and for Amsterdam the number of 2014 tweets fell between that of the national elections' 2010 and 2012 numbers (Table 6.1).

Most candidates have several hundred followers on Twitter, though the average number is considerably higher in Amsterdam. The typical candidate has about 400 followers. Only 6 of 1,500 local candidates had more than 5,000 followers (30 of the 500 national candidates topped that number in 2010), two of which were local list pullers and the other four "list pushers"—well-known people who are placed at the bottom of the list, all of them MPs or former MPs (e.g., Boris van der Ham, who had 50,000 followers).

Local politics is clearly still very much in the early-adopter to early-majority phase. This was also confirmed by the interviews with local campaigners and social-media managers, all of whom agreed

Table 6.1 Localc andidates'T wittera ctivity

	Amsterdam	Groningen	Almere	Nijmegen	Houten	Wageningen
Candidates on Twitter	62%	44%	41%	33%	30%	38%
Average daily number of tweets[a]	5.2	3.5	3.8	3.7	1.0	2.1
Median daily number of tweets[a]	2.2	1.3	1.7	1.3	0.6	0.9
Average number of followers	1,239	510	535	669	253	228
Median number of followers	417	192	265	256	85	183

[a] The number of tweets for Almere and Nijmegen was measured over the last seven days of the campaign. For the others, the last 20 days. This probably biased the average of the first two upwards, as also shown int het able.

that their social-media campaigns started too late. One, for instance, mentioned:

> *I was not a part of the core campaign team. That team was mainly thinking about posters, flyers, and logos, the traditional way of campaigning, when I took over the website from the previous webmaster, who quit his job. (...) I did not start a genuine social-media campaign (...) I had plenty of other things to do. (...) [Y]ou can't just start a social-media campaign a month before the elections.* (Inwonerspartij Toekomst Houten, social-media manager, 2014)

The social-media manager of one of the local branches of the conservative liberals (Groningen) was also very open about this:

> *We concluded that we started too late—especially regarding building our follower network.* (VVD Groningen, campaign leader, 2014)

Interparty Differences in Diffusion

Table 6.2 summarizes the quantitative diffusion of Twitter for the candidates, as well as the quantitative diffusion of Twitter and Facebook for the parties. Table 6.3 includes our indicators of the quality of the parties' Facebook posts. It is almost surprising how consistent the patterns are across municipalities, and how similar they are to our findings in chapter 4. One result deserving special note is that the interparty differences in Amsterdam are relatively small compared to those in the other municipalities, but even here the same patterns emerge.

Across the board, the three postmaterialist parties and two of the three big parties, namely the social democrats and the conservative liberals, have the highest number of candidates on social media. Of those parties, D66, GL, PvdA, and VVD have the more active candidates in terms of Twitter posts, and, in the four municipalities on which we have Facebook data, the Greens are persistently among the parties with the highest-quality accounts. As the postmaterialists do particularly well, the quantitative data suggest that motivation plays a role, but the overall relatively low quality of the usage hints at a lack of resources and expertise in some municipalities. This is most pronounced for the smallest of the postmaterialists, the Party for the Animals (PvdD), which had many candidates on social media, but presented low-quality Facebook posts.

The position of the Christian democrats (CDA) is intriguing. Nationally, it is the third largest party, and it even became the largest

Table 6.2 Percentage of candidates on Twitter and having party accounts

Party	Municipality						Overall[d]
	Amsterdam	Groningen	Almere	Nijmegen	Houten	Wageningen	
VVD	67	73	59	41	62	**58**	60
PvdA	74	50	57	43	53	52	55
CDA	34	49	18	33	52	39	38
D66	73	78	33	61	53	**52**	58
PVV			38[f]				38
SP	61	45	29	40		13	38
GL	87	65	62	45	32	52	57
CU		52	45		37	***64***	50
SGP					17		17
PvdD	67	71					69
Local "elderly"[a]	28[c]		15[c]	*0*[c]			14
Local 1[b]		46	44	24	75	14	33
Local 2[c]		25		12	21		
Overall[e]	62	44	41	33	30	38	

[a] These are not branches of 50Plus, but independent local parties for elderly. Amsterdam: Party of the Elderly; Almere: Almere Partij/Ouderen Partij Almere; Nijmegen: Verenigde Senioren Partij.
[b] Groningen: Stadspartij Groningen; Almere: Leefbaar Almere; Nijmegen: De Nijmeegse Fractie; Houten: Houten Anders!; Wageningen: Stadspartij Wageningen.
[c] Groningen: Student & Stad; Nijmegen: Gewoon Nijmegen; Houten: Inwonerspartij Toekomst Houten.
[d] Simple unweighted average of the municipal percentages.
[e] Percentage for all candidates pooled, thus weighted for the number of candidates listed by each party.
[f] There is Twitter account PVV_Almere, but it cannot be identified as being official and has no posts.

Note: **Bold** number indicate that the party has no local Twitter or Facebook account; ***bold italic*** numbers indicate that the party has neither a local Twitter nor local Facebook account.

local party nationwide after the 2014 elections (cf. Table 3.2). This is mainly because it dominates the rural areas, however, not the cities studied here. It should thus not be surprising that the party falls behind the parties discussed above in terms of number of candidates on Twitter: In cities the CDA is a small party (Table 6.2).[4] The Christian democrats do perform better than the remainder of the parties though, which is an indication that the central party leadership pushed them a bit. Indeed as one social-media campaigner of an independent local party put it:

I am also active on Twitter. And then suddenly two weeks before the elections a lot of these candidates pop up. (. . .) Some of them clearly just replicate the instructions they got from their party headquarters. They add the #CDA, but forget to add a space between their last word and the hashtag. I thought, my God, they just got instructions to add #CDA after each and every message they send out. (Inwonerspartij Toekomst Houten, social media manager, 2014)

Table 6.3 Facebook party accounts

Party	Quantity	Quality: Professionalism		Posts attempting to mobilize citizens (%)	Quality: Interactivity		Quality average rank (0–1)—the lower the rank the higher the quality
	Number of posts	Posts with substantive political information (%)	Post with pictures, video, or hyperlinks (%)		Likes and comments per post per 100 followers	Post where party reacts to comments (%)	
Amsterdam							
CDA	26	42	100	8	2.0	10	0.5
D66	37	41	100	11	2.5	19	0.5
GroenLinks	29	52	97	3	1.6	42	0.5
PvdA	46	41	100	37	1.2	13	0.5
PvdD	35	29	83	14	1.5	26	0.7
SP	23	48	100	17	2.6	7	0.4
VVD	28	57	100	21	0.6	0	0.4
Partij van de Ouderen	7	43	86	0	18.9	67	0.5
Groningen							
ChristenUnie	35	31	94	23	5.7	20	0.4
CDA	21	62	90	19	7.7	23	0.3
D66	35	31	97	23	2.1	9	0.5
GroenLinks	51	59	98	12	2.2	21	0.3
PvdA	18	50	18	22	4.2	0	0.6
PvdD	26	19	58	42	1.6	18	0.7
SP	35	46	91	6	3.6	11	0.6
VVD	8	50	100	13	1.8	0	0.5
Stadspartij Groningen	28	50	75	11	2.5	8	0.6
Student en Stad	53	72	94	0	1.0	0	0.6

Houten							
ChristenUnie	No account						1.0
CDA	34	29	100	24	8.2	0	0.4
D66	32	6	94	19	2.8	0	0.6
GroenLinks	24	33	100	25	8.2	0	0.3
PvdA	16	81	100	0	9.9	25	0.3
SGP	20	20	80	0	2.5	0	0.8
VVD	15	80	100	0	7.0	0	0.4
Inwonerspartij Toekomst Houten	9	78	100	13	12.5	60	0.2
Houten Anders!	13	69	92	8	5.8	50	0.5
Wageningen							
ChristenUnie	No account						0.8
CDA	37	49	92	3	1.6	0	0.3
D66	No account						0.8
GroenLinks	45	42	93	24	2.2	0	0.2
PvdA	13	31	92	23	8.7	33	0.2
SP	0	0	0	0	0.0	0	0.8
VVD	No account						0.8
Stadspartij Wageningen	10	50	90	0	18.3	86	0.2

Notes: Rank averages are calculated for each municipality separately.

Source: Jeroen Hellebrekers Bachelor thesis as part of VIRAL project. The scores are based on the posts between February 28, 2014, and March 20, 2014.

Most behind are the parties for elderly and the ultraorthodox Christian party SGP, which scores low on quality too. The other Christian party's local chapters (CU) are clearly not as advanced as the party is on the national level. Overall it falls in-between the leading parties and the laggards (see below). The independent local parties are generally among the parties left behind as well, but the diversity among these parties is also reflected in the quantity and quality of their social-media adoption.

Finally, the two populist parties are ahead of the smallest parties (PvdD and SGP) but generally behind the three bigger and the post-materialist parties. For instance, the PVV Almere has no Facebook account, and its candidates' Twitter adoption rate is below average. As in the national analyses, the SP shows some exceptionally active candidates, but at the same time it has a nonactive Facebook account in Wageningen, for instance. Moreover, with the exception of Amsterdam, their local posts are of a relatively low quality.

Interparty Equalization and Normalization

While the analyses of the municipalities clearly reflect our results from chapter 4, they do not merely present a simple repetition of the national analyses. The size of the municipality and the help of a mother party at the national level particularly seem to matter in this respect.

The training ground for many national politicians, Amsterdam is the municipality with the largest and most professional local party organizations. It also has the most widespread and most professional social-media use, which fits the resource dimension of our model (cf. supra). Moreover, social-media adoption rates of the Amsterdam parties are relatively similar, with only the elderly and Christians lagging behind. In the other municipalities, the independent local parties are generally behind. This is probably due to a lack of resources. One local campaigner of an independent party was very clear about the limited resources they had to work with compared to the other parties:

> [T]hat must have costs us about 600 Euros. So, not that much (...) look at the SP or PvdA or any other party, they simply have more manpower (...) parties that operate nationally simply have more budget available than we do. (Stadspartij Groningen, social media manager, 2014)

The qualitative pattern is more mixed, which might be related to whether or not a party had in-house expertise, but given that most independent local parties cut across socioeconomic cleavages (and are no clear-cut postmaterialists, cf. Boogers, 2010), this might be more incidental. Not surprisingly, several of the five interviewees responsible for social media had a background in communication. Moreover, the independent parties with which we did an interview did have a person responsible for social media, whereas not all local independent parties have one. It were also these parties[5] that showed high interactive behavior on social media and more so than the other independent parties. This suggests the importance of the right people in the right place. Similarly, the interviews with national social-media managers clearly showed that they tried to transfer their knowledge to the local level through visits, training, handouts, and handbooks (CU, social media manager, 2014; D66, social media webcare, 2014; GL, social media manager, 2013; PvdA, social media manager, 2013; PvdD, social media manager, 2014; VVD, social media manager, 2013). The VVD interviewee in Houten confirmed they had received training, and their candidates were also the most enthusiastic about social media of all parties in Houten (Table 6.2).

Furthermore, in terms of equalization and normalization, it matters whether one takes a national or local perspective. For instance, at the national level the Greens are a small party and their head start on, and strong usage of, social media implies equalization. In Amsterdam, Nijmegen, and Wageningen, GroenLinks is among the bigger parties. In the 2010 elections in Nijmegen, GroenLinks even was the largest party. Their lead in social-media usage implies normalization at the local level because they were powerful and not challenged by the introduction of social media as they also are very present on social media. Social media thus strengthened the party's offline dominance (though it constitutes equalization from the perspective of the national-level power balance).

Intraparty Equalization and Normalization

For the municipalities, fairly limited data is available on intraparty differences, but the numbers are quite clear. Of the 54 local list-pullers, only 6 had no Twitter account—89 percent did.[6] As for the other top local candidates (positions 2 to 6), we find that at least 3 candidates (60%–100%) on each list are on Twitter in 42 of 54 cases. Of the 12 parties that show lower numbers, 9 are orthodox Christian, elderly, or populist parties. Given that the overall adoption rate is well below 60 percent (see Table 6.2), the concentrated personalization we found in chapter 5 holds for the local elections too. The position of the local political leader seems strengthened by social media, particularly among the lagging parties. The list puller is among the most popular candidates in almost all cases. Yet this difference is clearly less pronounced than in national politics.[7]

However, as we also found in chapter 5, there are clear examples of lower-ranked candidates who have managed to build strong networks. Actively tweeting seems important: for the candidates in the six municipalities who are on Twitter ($n = 691$), we find a Pearson correlation between the (logged) numbers of followers and tweets of 0.57 ($p < 0.001$). This is considerably stronger than that among the national candidates in 2012 (0.46; $p < 0.001$), which clearly suggests that in local politics being an actively tweeting candidate is even more important in building a larger network. In terms of party affiliation, the candidates from the postmateralist (D66, GL) and non-Christian major parties (PvdA, VVD) again seem to be most strongly represented.

For instance, of all the candidates studies here, 21 have more than 3,000 followers; 12 from Amsterdam, 6 from Nijmegen, 2 from Groningen, and 1 from Houten. If we cross out the 4 local list pullers as well as the 7 national MPs or Senators, we are left with 4 Green candidates, 2 progressive liberals, and 1 each of the social democrats, conservative liberals, socialists, and an independent local party (5 from Nijmegen, 3 from Amsterdam, 1 from Almere, 1 from Houten). In other words, the candidates from the postmateralist parties seem to benefit most from social media in terms of intraparty equalization, at least in Amsterdam and particularly in Nijmegen.

As we have hardly any information on the quality of the candidates' social-media use, we should be careful here and not draw conclusions too strongly. And we particularly cannot perform systematic analyses with regard to gender and ethnic difference. Moreover, several of the higher-ranked candidates discussed above are working as communication staff members of the national parties and MPs, with large *national* networks, so they are not *locally* well-known politicians per se.[8] In short, the intraparty analyses in this section should be considered in the light of the book at large, and as such they fit the overall patterns of normalization with equalization for some, particularly for the postmaterialist candidates.

The Motivation-Resources Diffusion Model in Subnational Politics

The speed of the social-media diffusion at the local level is well behind that of national politics. In line with the logic of second-order political arenas, our expectation is thus confirmed on that front. It would be interesting to test whether the diffusion process is even less advanced in the smaller municipalities (around 20% of all Dutch municipalities have fewer than 15,000 inhabitants). In terms of which parties and politicians are leading, the patterns strongly resemble what we have seen in national politics (chapter 4), though there are some differences. The local branches of the Green parties are in the lead, even in 2014, and some of their lower-ranked candidates manage to be more visible than their own and other parties' list pullers using social media. As expected, the slower diffusion is to the Greens' advantage. Whether this is labeled equalization or normalization is a matter of which benchmark one takes: the Greens often were one of the bigger parties locally. However, even in those municipalities where the party is small—Houten and Almere—the party still clearly scored above average. In that sense it seems genuine equalization is taking place. Almere is instructive here: Even though the Greens have only 2 of the 39 seats, they were ahead of all the other parties in terms of social-media use.

At the other end of the spectrum, the orthodox Christian, elderly, and populist parties remain behind, with their top candidates least having to fear from social media as the other candidates in their party hardly use it (concentrated personalization). Particularly the bigger, nationally dominant parties show more diversity, which fits the slower pace of diffusion but also reflects differences between the power balance at the local and the national levels. Indeed, some of these parties might not be "bigger parties" locally, such as the Christian democrats in the municipalities we studied. However, all three are supported by expertise from their mother party, which probably explains why the Facebook pages of the local Christian democratic branches are of decent quality even though its relatively traditional local candidates have not yet embraced social media. The importance of national mother parties also explains why independent local parties tend to be lagging. Having a mother party offers additional resources (expertise, support, training, funding) and is clearly an asset.

European Politics

Below we focus on the supranational level: the Dutch parties and politicians in Europe's democratic arena, the European Parliament. In 2009, the Dutch population elected 25 of the total of 736 MEPs, and 26 of 766 in 2014. In the Netherlands, the elections are organized nationally (de facto functioning as one electoral district) and use the typical Dutch list-proportional system. One major exception is that the preference-vote threshold is not 25 percent but 10 percent of the electoral quota (about 18,000 votes). In the last three elections, turnout was 35–40 percent, whereas it was 75–80 percent in the recent national elections, reflecting the second-order status of European politics (Reif & Schmitt, 1980:9). The two major postmaterialist parties tend to do relatively well in these elections, as their electorate value European politics more. The Christian parties also tend to do well, because their electorates are more faithful voters (and because farmers, who mainly vote Christian democrats, tend to be well mobilized to protect the EU's Common Agricultural Policy).

Supranational versus National Diffusion

The early adoption of social media in European politics has largely taken place simultaneously with national politics: 14 percent of the candidates had a Twitter account in 2009,[9] which aligns with the diffusion at the national level, for which we found that one-third for the national parliamentary candidates had a Twitter account a year later. Moreover, most parties opened specific European Twitter accounts as early as 2008 or 2009. By 2015, Twitter is clearly ingrained in European politics. All European Parliament (EP) delegations, except the populist PVV, had their own Twitter account (see Table 6.4); traditional media organized Twitter debates for the 2014 elections (Europees Parlement Informatiebureau Nederland, 2014); all but one MEP have their own politically used Twitter account; MEPs have sent out several tweets a day on average; and more than 80 percent of candidates had a Facebook account and used it politically (see Table 6.5).

In terms of followers, the parties' accounts' networks are relatively small, but they use the national parties' accounts too. Regarding the individual politicians, the 2009 EP *candidates* averaged about 300 to 400 followers,[10] and in 2015 the *MEPs* had on average about 6,000 followers, with a median of 3,000. This median equals that of the MPs in 2012. Among the MEPs there were no politicians with more than 400,000 followers, none were as popular as Geert Wilders or Alexander Pechtold. On Facebook, the MEPs managed to build networks totaling 1,000 to 5,000 friends. All in all, it seems safe to conclude that the European political actors really invest in social media.

Table 6.4 Social media use in European politics per party

Party	Party: Facebook		Party: Twitter		Candidates on Twitter		
	Specific European Facebook account	Likes May 2015[b]	Specific European Twitter account	Followers May 2015[b]	% 2009 candidates on Twitter[a] (%)	Tweets per day 2009 campaign all candidates[a]	2009 Following party or candidate (x 100)[a]
VVD	No account	n.a.	Jan 2015	351	17	11	10
PvdA	June 2014	859	Mar 2009	2,874	31	3	11
CDA	No account	n.a.	Jan 2009	2,175	12	5	16
D66	WG account[c]	n.a.	Oct 2009	3,239	20	23	43
GL	WG account[c]	n.a.	Dec 2008	1,642	32	13	31
PvdD[d]	No account	n.a.	Nov 2014	286	7	1	2
CU/SGP	Apr 2014	1,644	(CU only) Oct 2009	1,029	5	0	0
SP	No account	n.a.	No account	n.a	3	0	0.5
PVV	No account	n.a.	Dec 2009	1,157	0	No accounts	No accounts

[a] Calculated from Verger, Hermans, & Sams (2013). Tweets are rounded to whole numbers based on Figure 1 in their study. The same holds for the followers.
[b] 2015 data refers to May 13.
[c] Both parties have EU working groups that have a FB account, this is not an account of the EP delegation.
[d] No seat after in 2009 elections.

Table 6.5 MEPs on social media (2015)[e]

MEP (Delegation leaders and (previous) list pullers in bold)	Twitter since	Number of follower[a]	Number of Tweets[a]	Facebook account or not	Number of friends/likes	Klout score[f]
Hans van Baalen (VVD)	May 2009	3,081	2	N	n.a.	49
Cora van Nieuwenhuizen-Wijbenga (VVD)	Nov 2009	5,908	5,263	Y	1,869	54
Jan Huitema (VVD)	May 2010	3,261	1,593	Y	1,380	52
Paul Tang (PvdA)	March 2009	6,296	4,585	Y	1,493	53
Agnes Jongerius (PvdA)	unknown	2,670	2,390	N	n.a.	54
Kati Piri (PvdA)	unknown	1,847	1,388	Y	unknown	52
Wim van de Camp (CDA)[c]	Jan 2009	14,400	13,600	Y	3,084	56
Esther de Lange (CDA)[c]	Apr 2009	6,352	3,696	Y	1127	52
Jeroen Lenaers (CDA)	Nov 2011	1,335	1,425	Y	575	46
Annie Schreijer-Pierik (CDA)	unknown	3,391	1,920	N	n.a.	51
Lambert van Nistelrooij (CDA)	Sept 2010	5,225	12,200	Y	1,244	53
Sophie in 't Veld (D66)	Jan 2009	22,600	26,900	Y	2,506	61
Marietje Schaake (D66)	Jan 2009	32,000	42,300	Y	5,571	66
Gerben Jan Gerbrandy (D66)	Apr 2009	6,294	6,133	Y	1,071	53
Matthijs van Miltenburg (D66)	unknown	1,377	2,342	Y	unknown	48
Bas Eickhout (GroenLinks)[d]	Jan 2009	10,900	22,600	Y	4,111	60
Judith Sargentini (GroenLinks)[d]	Jan 2009	8,345	12,100	Y	2,661	61

Anja Hazekamp (PvdD)	May 2010	2,259	2,887	Y	817	50
Peter van Dalen (ChristenUnie)[b]	Jan 2009	1,029	351	Y	789	50
Bas Belder (SGP)	Dec 2010	1,976	898	N	n.a.	47
Dennis de Jong (SP)	May 2009	2,648	3,318	Y	117	52
Anne-Marie Mineur (SP)	Nov 2009	1,941	41,000	Y	unknown	52
Marcel de Graaff (PVV)	Oct. 2010	1,947	4,884	Y	475	53
Vicky Maeijer (PVV)	Dec 2011	157	91	N	n.a.	42
Olaf Stuger (PVV)	Apr 2013	402	163	Y	1	45
Hans Jansen (PVV)	No	n.a.	n.a.	N	n.a.	n.a.

[a]May 13, 2015.
[b]This is on account with the party name, not his personal name, but it links directly to his personal website and is used as if a personal (though political) account
[c]Wim vd Camp was listpuller, but Esther de Lange became delegation leader.
[d]Bas Fickhout was list puller in 2015, but Judith Sargentini in 2009.
[e]Hans Jansen passed away in May 2015 and his position has not been filled yet. The second line of his website reads: "I do not have any Facebook or Twitter account."
[f]www.twitalyzer.com/ measured for the last five days (accessed on May 17, 2015).

Interparty Equalization and Normalization

The most important Facebook- and Twitter-adoption and activity figures (Tables 6.4 and 6.5) show the expected patterns. During the early-adoption phase back in 2009, D66 and the Greens were clearly ahead: they were the first to open accounts, had the biggest networks by far, tweeted the most, and were active both before *and* after their campaigns (Vergeer, Hermans, & Sams, 2013). They maintained this position until 2015 and for instance were the only party with all of its MEPs on Facebook and with 60+ Klout scores (which is a broad indication of high-quality use). At the same time, the bigger parties did partially close the gap. A major party's European social-media manager confirmed that it really was a matter of catching up. She started in 2012, and during the interview she told us that:

> *If I am completely honest, when I came here, there were social media accounts, Twitter and Facebook, on which something was posted only like once every three months or so.* (CDA, social media manager, 2014)

Using their financial resources, they provided training, an ad budget, and professional software.

Among the postmaterialist parties, the Party for the Animals was behind, but then they only won their first European seat in 2014, as was indicated by their social-media manager a few months after the elections:

> *[On social media] we provide information (...) about what the PvdD does. That is mainly focused on [the national] parliament. (...) we have expanded to Europe lately, but of course we remain a small party. We started in the lower house and Europe has been added to that now.* (PvdD, social media manager, 2014)

The two populist parties and the orthodox Christians were clearly lagging behind in 2009, with hardly any candidates on Twitter and often only opening accounts after the elections, if at all. Mostly the list pullers were the ones with an account, but for the PVV, this was not even the case, suggesting that the overall party leader—Geert Wilders—wanted to control the European communication too. In 2015, the candidates of these parties still used Facebook less than their colleagues from other parties; their networks were small; they either had no Twitter account or MEPs signed up to Twitter very late; and both SP and PVV were the only parties that did not participate in the Twitter debates (Europees Parlement Informatiebureau Nederland, 2014). Of these parties, the CU stands out slightly and seems to have a more developed Facebook strategy, while the SP has one very active candidate (Table 6.5).

Overall, the data on European politics are very similar to that of the national interparty analysis (Chapter 4): equalization in the first phase for postmaterialist parties, bigger parties catching up in the second phase (normalization), and the populist and non-postmaterialist smaller parties lagging behind (qualitatively) in the last phase of diffusion. The latter are the "losers of digitalization." At the same time though, the patterns are not identical. If anything, the postmaterialists' advantage and the populists' disadvantage seem more pronounced, and the Christian democrats seem to do particularly well among the bigger parties. The latter is not that surprising: The Christian democrats are part of the largest European party (EPP) and thus received their support (CDA, 2014).

Intraparty Relationships: Some Diffuse Personalization

In 2009, only 14 percent of all EP candidates were active on Twitter, but most parties' list pullers had a Twitter account (only two did not). None of the PVV candidates (!) had signed up for Twitter, and only one of the ten orthodox Christians candidates was online, but not the list puller. In the other populist party, the SP, only the list puller was on Twitter. He signed up about a month before the elections (see also Vergeer, Hermans, & Sams, 2013). Most other candidates from the two postmaterialist parties and the social democrats were on Twitter. The candidates for the two postmaterialist parties were the most active and had the largest networks (see Table 6.4).

By 2015, (former) political leaders generally had the most followers and were among the most active candidates. Though six of the eight former or current European political leaders are in the follower top ten, flanking them there are highly active candidates from the postmaterialist parties, who sometimes are even ahead of them in the top ten. At the other extreme, the MEPs of the populist PVV lag behind their list puller and the normalization process is particularly strong in that party. In terms of followers, the list puller of the other populist party, the SP, is also well ahead of the other socialist MEP, even though the latter was far more active on Twitter (Table 6.5).

For the European arena, data on professionalism is scarce, not to mention that professionalism is a rather complex concept in European politics. For instance, the CDA's social-media manager highlighted the issue of language:

> *I advised them to tweet in English during hearings, because there is more media attention then. Also because they have certain responsibilities [in the European Parliament]. (...) Wim van de Camp is transportation policy coordinator for the [cross-national] European EVP delegation, so I asked him to send a tweet in English every now and then.* (CDA, social media manager, 2014)

One way we could explore the interactivity and engagement of MEPs on social media was retrieving a recent Klout score and comparing these scores to the MEPs' numbers of followers and tweets.[11] If the score is relatively high, this indicates more engagement and higher-quality use of social media. For instance, the fact that Marietje Schaake from D66 does better than her political leader is not surprising given her larger network and greater tweet activity. Most of the female MEPs perform well compared to their leaders: Agnes Jongerius (PvdA) has a relatively high Klout score, indicating that she tweets qualitatively well, as does Judith Sargentini (GL). In other words, several MEPs manage to use social media at a high-quality level, which may equalize intraparty relationships. And once again the postmaterialist MEPs present the most pronounced examples of this pattern.

Intraparty Relationships: Gender Inequalities

In terms of underrepresentation, only one MEP in 2015 can be considered to have an ethnic-minority background, and as the study of Vergeer, Hermans, and Sams (2011) does not include information on ethnicity little can be said in this respect. However, Vergeer, Hermans, and Sams (2011) did include people's sex and found that women were substantially more likely to have an account and to have started tweeting earlier.[12] However, after controlling for party affiliation and other factors, women were found to have similar or slightly lower tweet counts, interactivity (@-mentions), and numbers of followers. In other words, they found that women seemed more eager to join social media, but this was mainly due to party affiliation, and their activity was slightly lower or identical to that of their male counterparts. While similar, this pattern of overrepresentation was slightly stronger for women in the 2009 EP elections than for those in the 2010 national elections.[13]

For the 26 MEPs in 2015, the differences in social-media adoption rates have apparently evaporated, as almost everybody has social-media accounts. Regarding the activity and the quality of usage we do find important gender differences: on average women have about twice as many followers and post roughly 2.5 tweets for every tweet by a male MEP, and the examples of MEPs doing particularly well presented earlier were all women. Party affiliation also seems to play a role: (gender-balanced) postmaterialists clearly do better and (mostly male) populists do worse, but these factors do not explain all differences. Even most female MEPs of the social democratic, Christian democratic, socialist and conservative liberal parties do better than their male counterparts (cf. Table 6.5). Yet, as we ultimately only have a fairly low number of cases to work with (26), though these patterns are suggestive, they may be an artifact of aggregating several exceptional politicians.

The Motivation-Resources Diffusion Model in Supranational Politics

Overall, social-media adoption and usage are on par with the diffusion in the national arena, but European politics stays behind in network size. This should not come as a surprise given the lower degree of media attention. Moreover, the number of politicians who use social media at a high-quality level seems relatively high. This echoes the fact that the more postmaterialist parties are relatively more powerful in Europe and have a more Europe-oriented and tech-savvy electorate.

In terms of inter- and intraparty equalization and normalization, the analyses in this chapter by and large confirm our previous findings. The major differences are that the patterns tend to be stronger and that social media appear more beneficial to postmaterialist parties, previously disadvantaged candidates, and female candidates, and less so for their populist and traditional counterparts, who experience concentrated personalization and generally lag behind even more than in the national level analyses. All in all then, the quick-normalization expectation seems to be corroborated and the difference between the local and European level is striking: European social-media use is more similar to national social-media use, though there are some more hints of equalization.

Comparative Perspectives

Social-media use in local and European arenas is rarely studied (Larsson, 2013),[14] because the focus in the literature mainly falls on the more salient political arenas. Two of the most prominent studies on European politics have already been discussed above, as they concern the Netherlands (Vergeer, Hermans, & Sams, 2011, 2013), which leaves us with very little comparative material as far as we are aware: one on Belgium (Jacobs et al., 2015), one on Switzerland (Klinger, Rösli, & Jarren, 2015), and one on Sweden (Larsson & Svensson, 2014).

In Belgium, the number of candidates with a Twitter account is higher in the geographically larger arenas: more EP candidates have a Twitter account (28.6%) than federal candidates (21.3%) and more of the latter have accounts than regional candidates (18.4%). The number of tweets posted also follows this order, and in all cases the list pullers are far more popular than their colleagues (Jacobs et al., 2015:15, 17). The fact that the European social-media adoption rate is higher than the federal one deviates from our results, yet the underlying mechanism could be similar. The Belgian Green parties—the main postmaterialists—scored some 11 percent of the votes on average in the 2014 European elections, while in the federal election they averaged about 7 percent of the vote share. This stronger presence of postmaterialists at the European level could partly explain the higher overall adoption rates of social media in the European elections, but a more in-depth study is needed to confirm whether this really is the case.

The Swiss study by Klinger, Rösli, and Jarren (2015) examines the use of social media by *municipalities themselves*. Yet their main findings are actually in line with what we found in our interparty analysis: social-media use is less widespread at the local level and the choice not to adopt social-media strategies is largely dependent on lower motivations due to the relatively low demand of such tools by citizens and the lack of financial resources (Klinger, Rösli, & Jarren, 2015:1934). The Swedish study also deals with the social-media accounts of Swedish municipalities themselves, not those of political actors (Larsson, 2013). It does however provide some relevant information: again it is the larger (more resource-rich) municipalities as well as the more left-wing (social-democratic) oriented municipalities that have accounts on Facebook and Twitter. Both these elements support the reasoning we propose in this chapter.

If anything, these three studies seem to confirm the core mechanisms we proposed based on the Dutch case study. However, they help little in understanding the differences between first-order and second-order political arenas—a main goal of this chapter—as simply not enough comparative interlevel data is available. This backs up Larsson's calls for more analyses of local politics (Larsson, 2013; Larsson & Svensson,

2014) and allows us to add that explicit comparisons with national-level politics would be particularly useful, as national party headquarters and national political differences seem to heavily influence the sub- and supranational quantitative and qualitative diffusion.

Conclusion

In this chapter we focused on a specific aspect of our second research subquestion—How does political context affect the impact of social media on the power-balance position of parties and politicians?— namely the type of political arena (first-order arena versus second-order arena [Reif & Schmitt, 1980]), which has important implications for the saliency of elections and the geographic distance between politicians and citizens.

Differential Diffusion: Geographic Distance and Professionalism

At the European level, the political diffusion of social media equaled that at the national level, whereas at the local level it was clearly behind, but least so in the Dutch capital, its most populous city. If only geographic distance and a lack of saliency were leading in social-media adoption, the European adoption should be ahead of national politics (see chapter 2), but that was not the case. Professionalism seems a more important explanation: European and national politics are equally professional and organized from the national headquarters; local politics is most professional in the capital, where it has the best funding and often functions as a stepping stone toward national politics. In the other municipalities, diffusion is clearly behind, as expected.

Local-level politics was slow to adopt social media and diffusion took more time. However, the interparty differences are not markedly stronger in subnational than in national politics, which was our initial expectation. Yet let us not forget that we examined the most likely cases only. In such settings similarities to the national level are more probable. We still discovered some intriguing patterns though. One concerns the efforts of national parties' headquarters: the parties tried to train their local branches and the local parties clearly faced an uphill struggle. This is also reflected in the fact that the biggest local parties are not always the strongest on social media. In this light, taking an equalization-normalization perspective becomes quite complex for the local level, because national equalization might imply normalization at the local level (cf. the greens in the city of Nijmegen). Our analyses suggest that the power-balance shifts at the local level can best be understood by analyzing the motivational and resource elements in light of the national power inequalities and then looking at the implications of these at the local level.

As mentioned above, the European patterns might point in the same direction, but the strength of these patterns is somewhat different. Contrary to our initial expectations, the positions of the postmaterialist and lagging parties seem more extreme in the European arena, even though the overall adoption rates are high. This can be ascribed to the lasting head start of the postmaterialist parties and their relatively strong position in the EP, as well as the role national parties play in coordinating social-media activities. However, the parties that lag behind are seriously behind, so far that we can even call them "losers of digitalization."

Equalization Is Normalization. Similar Patterns, Different Conclusions?

Whether we focus on political arenas that are social-media saturated (national and European) or on arenas in which the overall diffusion of social media is still far from complete, the patterns are very robust. Across arenas, the postmaterialist greens take the lead in the first diffusion phases and they remain ahead. Among their ranks also non-top candidates especially stand out. They have more intrinsic motivation and expertise available, even if financial resources are limited. Similarly, the traditional (elderly, ultraorthodox) and populist parties lag behind and their list pullers remain the most dominant on social media. The three bigger parties and the postmaterialist D66 show somewhat more divergent positions, depending on the diffusion phase and the political arena. This is illustrated by the Christian democrats, who do have the resources and stimulate high-quality social-media usage at the European level, though their more traditional candidates need more "nudging" than those of other parties.

The exceptions actually illustrate the underlying mechanisms of our motivation-resources-based diffusion model. In the second-order arenas, the Party for the Animals, PvdD, is lagging behind more than at the national level where it is among the large group of leading parties. This is not due to a lack of "postmaterialist motivation," but the tiny party simply does not have the resources to be on top of everything. Similarly, the independent local parties cannot benefit from the support of parties' national headquarters, which explains their general laggardly position. The fact that some independent parties stand out can be explained by the same mechanism that explains the divergent position of the small ChristenUnie (see chapter 4): sometimes there are social-media fans among a party's volunteers and these manage to lift those parties up.

Future Research: Some Suggestions

Future analyses could focus on (a) smaller towns—for which we would expect very low quantitative diffusion levels; (b) the way the dominant parties in a municipality can explain diffusion—whereby we would expect low adoption in traditional Bible Belt municipalities; (c) the provincial level—for which a saliency and geographies reasoning would predict adoption levels between the local and national/European ones; and (d) the differences between European and national diffusion—our study suggesting that the national and European arenas do not differ much.

PART III

―――――

The Transformative Impact of Social Media

CHAPTER 7

Do Social Media Help Win Elections?

Introduction

So far we have focused on which candidates and parties adopted and made professional use of social media such as Twitter and Facebook. We examined to what extent social media level the playing field and found that social media allow postmaterialist parties to communicate as if they were bigger parties. We also found that some candidates fully exploit the opportunities of social media while others do not. Especially female politicians seem to do well. But does such a (partially) leveled playing field also yield electoral benefits? Do social media merely change the communicative power balance, not the electoral one? If so, that would significantly undermine the equalization argument. Indeed as Gibson and McAllister (2014:2) rightly observe, "if smaller parties are capitalizing on their digital campaign efforts but not gaining any inroads into popular support, then it becomes difficult to see how this is leading to a rebalancing of power within the system."

The previous chapter referred to numbers of followers and preference votes, pulling in the demand side of politics, but we did not systematically assess the electoral impact of the social-media diffusion in politics. In his chapter we examine whether social media use yields electoral benefits, first outlining why one could expect social media use to result in more votes. Specifically we will link three of the aforementioned social-media opportunities (advertisement, human contact, target group) to a potential *direct effect* of social media. The fourth and last opportunity (salon debate), however, could spark an *indirect effect*.

After that brief theoretical discussion, we will embark on a statistical analysis examining the effect of social-media presence and use on the votes candidates received in the 2010 and 2012 elections.[1] An analysis of parties' vote shares is currently impossible: too many variables, too few parties. For instance, we would not dare ascribe the Green party's increased vote share in the 2010 election (+2.1 percentage points) to the party being ahead of all the other parties on social media. Moreover, though the party was still in the lead on social media in 2012, it collapsed almost completely (–4.4 percentage points) to an all-time low of 2.3 percent of the vote in that year. Too many other and election-specific factors are at play.[2] The precise effect of social-media use for parties is thus extremely difficult to assess, or it would require large comparative databases and experimental research templates not currently available. At the same time, the intraparty effects provide a crucial first test of whether social media substantively influence voting behavior, and this can be examined.[3]

We will examine whether being active on Twitter correlates with more preference votes in general and see whether this effect has changed between 2010 and 2012. First of all, there are strong indications of social media having a modest impact on the candidate a voter casts her/his vote on, and by 2012 this effect becomes more widespread, suggesting a modest (intraparty) equalization effect of typical Twitter behavior. In combination with previous results (chapter 5), this also suggests that for some more professional candidates social media can provide that little extra push they need for a preference seat. We will study whether the effect differs for underrepresented groups, particularly women.[4] It seems that female candidates benefit more from social-media use than their male counterparts; in that respect a modest gender equalization seems to take place, as also found in previous chapters (chapters 5 and 6).

After the empirical analyses for the Dutch case, we can examine our findings from a comparative perspective. While most studies still examine the effect of Web 1.0 features such as personal websites, more and more studies on social media are coming out. Most of these find a small but significant effect of social-media use, just as we did. Finally, we summarize the main findings.

Why Social Media May Matter: Direct and Indirect Effects

In this book we have used four opportunities to guide our analysis of the impact of social media, all of which are also useful in analyzing the electoral effect of social media. In the electoral context, it is worthwhile to make an additional distinction between the direct and indirect effects of social media. On the one hand, the *direct effects* label refers to effects stemming from direct contact between the voter and the politician. *Indirect effects*, on the other hand, conceptualize the effect of social media as stemming from a multistep flow in which voters are not directly influenced by politicians, but by journalists, opinion leaders, and peers who follow that politician (cf. Bond et al., 2012).

Direct Effects: Reaching Out to the Voters

1. *Advertisement effect.* In its most optimistic form, one can expect that even minimal social-media use has an effect on preference voting. It showcases the candidate as being modern and allows them to add details about hobbies, interests, political views, sociodemographic characteristics, and the like. It facilitates "personality-centered" campaigns (Vergeer, Hermans, & Sams, 2013). In this sense social media function as the equivalent of a campaign poster or flyer, with the bonus of adding a halo of innovation and creativity. Especially during the early adoption phase, simply being present on social media can serve as a means to distinguish oneself and appear modern. In the later phases of the diffusion process, the effect is probably reversed: the few candidates who do not have an account look old fashioned. In these later phases, more professional use is probably needed to stand out from the crowd.

2. *Human-contact effect.* A particularly potent use of social media may be engaging directly with citizens, as social media enable and even stimulate two-way communication. Sanne Kruikemeier (2014:133) clearly outlines why interactive use of social media can yield a vote bonus:

> [I]nteractivity induces social presence, which consequently leads to higher intention to vote for candidates. Social presence is the extent to which an individual feels that another communicator (in this case the candidate) is present and there is an opportunity to engage in an actual conversation.

What is important here is that the candidate does not necessarily need to answer each and every message but rather show that they are approachable. Indeed the candidate needs to demonstrate that if a voter wants to ask a question, they will get an answer.[5] Conversely, candidates like Geert Wilders (PVV), who never answer questions, signal that they merely use social media to send out their message and do not listen. Such "pretence of presence" (Crawford, 2009:530) is unlikely to trigger a direct effect on voting. The human-contact effect may well remain strong in the widespread phase of social-media diffusion, partly because significant differences exist between parties and candidates in this respect, as we have shown in chapter 4.

3. *Target-group effect.* Social media also allow politicians to establish contacts with certain niche groups that are hard to reach via traditional media. For instance, Twitter does not require a physical or email address and is not geographically confined (Utz, 2009:240). Similarly, Facebook allows the user to target Facebook ads to specific age and social groups, and so on. As we mentioned earlier, several Dutch MPs and parties crafted policies targeting groups that might have been hard (and expensive) to reach otherwise. This effect can also be expected to remain present in the widespread diffusion phase.

Overall, we should not expect a one-to-one translation of following a politician to voting for that politician, partly because some followers will actually follow a candidate because they are critical of them. Nevertheless and on balance, a positive impact can be expected via these three mechanisms. Indeed it seems safe to assume that a number of followers will support the political ideology of the candidate. Among these "receptive followers", the advertisement, human-contact, and target-group effects suggest a (modestly) higher likelihood of voting for candidates that voters follow, and we should thus expect more votes for candidates who have more followers and post more messages.

Indirect Effects: Reaching Out to Opinion Leaders, Journalists, and Friends of Friends

4. *Salon-debate effects.* Some have argued that it is mainly opinion leaders and nonrepresentative citizens who populate social media (and Twitter in particular). Throughout the book we have shown several examples of social-media managers stressing that Twitter users are not a representative subset of the broader population.[6] However, there are several reasons why social media can have an impact even though many "ordinary citizens" are not on Twitter or do not follow a particular candidate. Each of these reasons starts from the contention that the effect of social media is mainly indirect, following a multistep flow.

As the social-media manager of the social democrats (PvdA) puts it, social media are "primarily used by 'movers' and 'shakers', people who are opinion leaders" (PvdA, social media manager, 2013). In that sense, social media are like the French "salons," a foreshadowing of what public opinion is to become (Herbst, 2011:95). Controlling or influencing social media "movers" and "shakers" can yield an electoral bonus, as journalists often scour Twitter to build news stories around tweets by prominent politicians or inflammatory tweets by less prominent ones (Peterson, 2012:432; see also chapter 3), thereby increasing the media exposure for candidates on Twitter. Similarly, opinion leaders and activists often search Twitter to find lines of argumentation and images that might strengthen their preferred candidates' case or undermine their opponents' (see Box 7.1). This connection to traditional media is also mentioned by many candidates and parties as a motivation to use Twitter (see chapters 3 and 5).

Box 7.1 An example of the use of infographics to influence opinion leaders

▶ "The best health care system of Europe"

▶ VVD Logo and slogan "VVD sails ahead"

▶ Accompanying text, linking graphic to television debate: "Hey Roemer, pay attention! We have the best health care system of Europe"

On the March 17, 2015, the day before the Provincial elections, one of the spokesmen of the conservative liberals tweeted an infographic at the exact moment when the party leader of the socialists, Emile Roemer, was criticizing the quality of the Dutch health care system in a television debate.

Formulating Context-Specific Expectations

As we analyze candidate data (and the party vote shares candidates receive), we focus on the combined effect of the four opportunities, each predicting a straightforward positive impact of social-media use on the number of preference votes a candidate receives (Expectation 7.1).

For now, we are unable to distinguish the four effects, but some other pieces of information inform our expectation about the way the impact of social media might change over the diffusion process. In the two years between the 2010 and 2012 elections, Twitter and Facebook became more integrated in Dutch society, and journalists and opinion leaders started using them more as an important source of information (see chapter 3). According to the Dutch election survey 2012, for instance, 3 of every 100 voters remember being contacted via social media about politics by politicians or people they knew.[7] Moreover, between 2010 and 2012, digital communication—combining Web 1.0 and Web 2.0—clearly became more important for Dutch voters as a political tool (see Figure 7.1).[8]

Additionally, candidates and parties became more proficient in using Twitter, partly through trainings commissioned by party headquarters (see previous chapters). This can be expected to increase the effectiveness of their social-media use, which for the lower ranked candidates implies that at least some professionalism is obtained. All of this

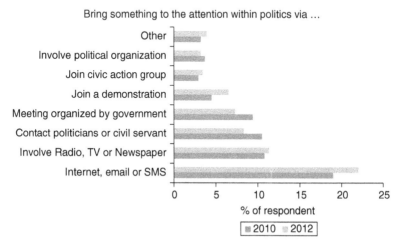

Figure 7.1   Means to influence politics used by the people.
Source: NKO,2 012.

suggests that in 2010 only candidates who were "naturals" competed for the scarce social-media votes. By 2012, a far more leveled playing field with more "regular voters" can be expected to have resulted in a more flattened-out and widespread effect (Expectation 7.2).

Finally, as discussed in chapter 5, women's underrepresentation in politics suggests that for women social media are a particularly interesting way to communicate, and the target-group and human-contact mechanisms might particularly help them to mobilize women (and men) to vote for them. Also, considering that party organizations are often gender biased (see Leijenaar, 1997), the salon-debate mechanism suggests that women have most to gain from social media. Voter wise, disaggregating the data in Figure 7.1 by sex shows that women's general political participation is almost four percentage point lower, but actually (slightly) *higher* for political activities on the web and on social media. In sum, there are reasons to expect that women benefit more from activities on social media (Expectation 7.3).

> Exp. (7.1) Candidates who use social media receive more preference votes.
>
> Exp. (7.2) In a context of widespread social-media use, social media still have an effect on the number of preference votes, but more candidates benefit (though to a smaller extent).
>
> Exp. (7.3) Female candidates who use social media receive more preference votes than their male counterparts.

Empirical Analysis, Part I: Are Not All Candidates Equal?

Table 7.1 shows the main results of an OLS regression analysis zooming in on the effect of Twitter[9] controlled for "the usual suspects"—variables that have been found to influence candidates' electoral success.[10] We included the number of followers as the reach obviously determines the potential impact of Twitter use. However, as merely having a popular social-media account without using it is unlikely to do the trick,[11] we also included the number of tweets sent out. As Twitter use is most likely to be successful when candidates both use social media *and* have a decent number of followers, we also included an interaction term. This term is the most important when assessing the effect of Twitter, as it is the most likely to measure a real effect. Lastly, we included the quadratic term to test whether sending out more tweets is electorally beneficial or not. It may well be that the added value of each additional tweet may decrease or that it may even lead to annoyance and negative effects (overloading one's followers with tweets). These analyses

Table 7.1 Candidate-levela nalysis

	Model 1: Early adopters (2010)	Model 2: Widespread use (2012)
Twitter followers (per 1,000)	0.003	−0.033**
Tweets per day	0.035	0.016**
Tweets per day* Followers (per 1,000)	0.007***	0.001*
Tweets per day [quadratic term]	−0.003*	−0.977E-4**
Adjusted R^2	0.617	0.371
N	159	383

*= $p < 0.05$; ** = $p < 0.01$; *** = $p<0$ .001.

Notes: The dependent variable measures preference vote shares per candidate per party; the theoretical maximum value is one, which refers to the case where a candidate received all the votes cast for that party. All models included controls for "the usual suspects." Coefficient are unstandardized B-coefficient. Dataset only includes candidates who had a Twitter account. List-pullers and first female candidates were excluded, but robustness checks show that including them yields in similar findings (see Jacobs and Spierings, 2015). For more technical details, see Jacobs and Spierings (2014).

Source: Jacobs & Spierings, 2014.

will show the general average effects of average social media use by candidates. In other words, we do not take the quality of posting into account. Also, list pullers and the first women on the list are excluded from the analyses because they are extreme outliers whose number of followers and preference votes are clearly largely due to their position, which would introduce a very strong positive bias to the analyses and lead to spurious relationships.

The results in Table 7.1 seem to corroborate Expectation 7.2, indicating that in both 2010 and 2012 there was an effect of Twitter use at large. The effect is modest though. When we calculate the equivalent in absolute terms it turns out that, *ceteris paribus*, average candidates from average parties with an average number of followers and typical tweet behavior receive some 200 to 500 extra preference votes (Spierings & Jacobs, 2014). Given that candidates need to cross a preference-vote threshold of around 16,000 votes, only using Twitter will not do the trick. Twitter campaigning will only work in combination with other campaign instruments. However, it might just tip the balance for a candidate who is close to the preference-vote threshold but low on the list. In fact, during the 2010 general elections, two candidates would not have gotten a seat without their preference votes, and only 250 votes fewer would have meant not becoming an MP: Sabine Uitslag (CDA, 15.933 preference votes) and Pia Dijkstra (D66, 15.705 preference votes).[12] Both actively used Twitter during the campaign. In short, though there is a modest effect of typical Twitter use, more can be expected of Twitter use that utilizes the specific opportunities defined above and in chapter 2, although this might leave fewer resources for other campaign activities. This modest (and for some potentially larger) effect can surely make or break the careers of individual candidates—at least in the short term.

The results in Table 7.1 also indicate that citizens may grow tired of endless streams of tweets: both in 2010 and 2012, the quadratic term was negative and significant. In 2010, however, it was 30 times as high as in 2012. In fact, in 2012, the effect of the quadratic term was so small that it only materialized when candidates sent out extraordinarily high numbers of tweets per day. Figure 7.2 illustrates the difference

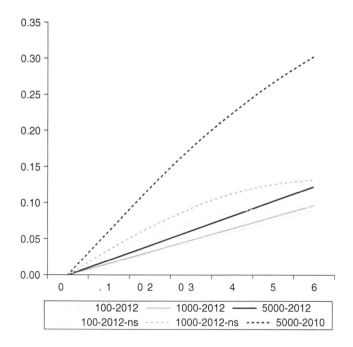

Figure 7.2   Estimated effects of tweets and followers on the vote share (2010 and 2012).

Note 1: "ns" = not significant.

Note 2: 2010 estimates depicted by dashed lines, 2012 estimates by continuous lines. The different lines depict estimates for the different numbers of followers. The X-axis shows the estimated effect as influenced by different average numbers of tweets per day. We limited the Y-axis at 6 tweets per day, as the number of cases quickly drops from that point onwards. An average candidate tweeted 1.9 times a day in 2010 and 2.7 times a day in 2012. Candidates who tweeted both in 2010 and 2012 sent out 5.5 messages a day (all median values). See also: Jacobs & Spierings, 2014.

in effects: While the dashed lines representing the 2010 effect clearly show a curve, the 2012 continuous lines are virtually straight. The latter are also far closer to each other. In fact, the 2010 effect only becomes significant at approximately 2,000 followers, indicating that only the popular candidates and social-media innovators, the "natural tweeters," benefited from Twitter use consistently. For candidates with smaller networks, the effects were more modest, more diverse, and less certain. In 2012, the effect was more leveled out, which is largely in line with Expectation 7.2. This indicates that lesser-known candidates virtually had as much to gain from Twitter use as their well-known counterparts. How to explain this? As we saw earlier in chapter 3, by 2012 most parties had started providing professional support to their candidates and the overall quality of the tweets was most likely higher. At the same time, Twitter became more widespread in society, which means that more diverse groups of citizens were present on the medium. There were more votes to win and more candidates were capable of winning something.

Empirical Analysis, Part II: Gender Equalization?

In chapter 5, we observed that female candidates were more likely to have Twitter and Facebook accounts than their male counterparts (though men did have more followers, probably because they often also have more prominent positions). Women seem to be keener on using social media, but it remains to be seen whether female candidates also attract more votes by using such media. To examine this topic empirically we carried out OLS regression analyses similar to the ones in Table 7.1, but split the file by sex (Tables 7.2 and 7.3).[13] In the 2010 election, fewer candidates were on Twitter, which means results should be treated with caution, particularly because the models include many control variables.

In 2010, we find that tweeting and posting messages had a significant effect on the preference-vote shares of male candidates, but posting tweets had less clear or less consistent added value for women. Since the number of followers did have a positive effect, either only well-known women were on Twitter or it was enough for women to open an account and show they were "modern" (cf. the advertisement effect). Either way, taking the median number of followers and tweets per day and plugging them in the regression equation for women and men shows that the effects for both sexes are rather similar (*ceteris paribus*).

In 2012, the situation is different. The size of the regression coefficient of the interaction term is 0.0014 for men: a very modest effect.

Table 7.2 Candidate-levela nalysis(meno nly)

	Model 1: Early adopters (2010)	Model 2: Widespread use (2012)
Twitter followers (per 1,000)	−0.024	−0.036**
Tweets per day	0.055	−0.006
Tweets per day* Followers (per 1,000)	0.010*	0.001**
Tweets per day [quadratic term]	−0.005*	3.935E-5
Adjusted R^2	0.635	0.371
N	100	250

*= $p < 0.05$; ** = $p < 0.01$; *** = $p < 0.001$

Notes: For technical details, see the notes accompanying Table 7.1.

Table 7.3 Candidate-levela nalysis(womeno nly)

	Model 3: Early adopters (2010)	Model 4: Widespread use (2012)
Twitter followers (per 1,000)	0.027*	−0.112***
Tweets per day	0.015	0.047***
Tweets per day* Followers (per 1,000)	0.003	0.002*
Tweets per day [quadratic term]	−0.002	0.000***
Adjusted R^2	0.535	0.528
N	59	133

*= $p < 0.05$; ** = $p < 0.01$; *** = $p < 0.001$.

Notes: For technical details, see the notes accompanying Table 7.1.

However, in the all-female analysis, this coefficient is almost double that size (0.0024). In 2012, tweeting was important for both men and women, but women benefited more than men—an indication of gender equalization.[14] Expectation 7.3 thus seems to be corroborated. Surely these findings provide no definitive answers, as especially in 2010 the number of women is quite low, but they are suggestive nonetheless.

Comparative Perspective

Above we found that, during the widespread diffusion phase (2012), female and lesser-known candidates did benefit somewhat from building a social-media audience and posting regularly. It seems that real-life campaigners (to some extent rightly) use social media in the hopes that it will deliver more votes, as is well illustrated by the fact that social-media posting skyrockets when elections draw near and plummets again when the votes have been counted. This is particularly true for the non-postmaterialist candidates (see for instance Kruikemeier, 2014; Vergeer, Hermans, & Sams, 2013).

Given this expectation among politicians and campaigners, one would also expect that scholars are keen to investigate the possible electoral effect of social media. Yet most research examining the connection between social-media use and votes does so in order to forecast election outcomes.[15] In fact, in his extensive literature review of research on Twitter, Andreas Jungherr (2014:49) observes that "[t]he effects of the use of Twitter by politicians are not really well understood and, somewhat surprisingly, not intensively researched." For other social-media types the situation is not better. However, the number of studies on the topic is fortunately on the rise overall, and when all the pieces are put together it turns out that the results are surprisingly consistent across contexts.

Compared the Netherlands to Other Western Democracies

The pioneer studies on the Internet and voting mainly examined the electoral effect of technological innovations in Anglo-Saxon countries. Indeed, the first study by D'Alessio (1997) examined whether candidates in the 1996 US national election benefited from having a website (Web 1.0), a study that was replicated later on by Wagner and Gainous (2009) in the context of the 2006 congressional elections. Afterwards, a string of papers by Gibson and McAlister (2006, 2011, 2014) examined the effect of Web 1.0 and Web 2.0 tools on the Australian 2004, 2007, and 2010 elections. All these studies find positive electoral effects either for Web 1.0 features (D'Alessio, 1997; Gibson & McAllister, 2006; Wagner & Gainous, 2009) or for Web-2.0 tools (Gibson & McAllister, 2011, 2014).

More recently, scholars have started to analyze the electoral effect of Web-1.0 and Web-2.0 communication tools in more proportional political systems. The impact of personal websites was studied in Ireland (Sudulich & Wall, 2010) and Poland (Koc-Michalska et al., 2014), and these studies found positive effects.[16] Regarding the impact of social-media campaigning in Belgium (Jacobs et al., 2015), Denmark (Hansen & Kosiara-Pedersen, 2014), Ireland (Suiter, 2015), and Italy (Ceron & D'Adda, 2015), all but one study found a (modest) positive effect of social media.

All in all, only 1 out of 14 studies finds no significant effect of Web 1.0 and Web 2.0 campaigning. The other studies exhibit results mirroring the ones on the Netherlands for the 2010 and the 2012 national elections (e.g., Jacobs & Spierings, 2014; Kruikemeier, 2014; Spierings & Jacobs, 2014).[17] Moreover, while earlier studies were often plagued by the lack of proper controls (e.g., for the overall professional level of the campaign and for media exposure), recent studies have become ever more sophisticated; typically even more sophisticated than more traditional preference-vote studies predating or excluding online media (cf. Lutz, 2010; Wauters, Weekers, & Maddens, 2010).[18]

The United States: Obama Leads the Way

That American scholars carried out the first studies should not come as a surprise: online campaign tools have typically found their first applications in the United States, where stakes are high and financial resources abundant. D'Alessio (1997)—see above—was the first, finding that candidates who had a website scored significantly higher than those without one. Research by Wagner and Gainous (2009:502) further supported this and found that online campaigning was indeed "a significant predictor of the total votes candidates garnered in the 2006 congressional elections, even when controlling for variables such as funding, incumbency and experience."

While personal websites seemingly still had an effect on vote shares, Democratic candidates had by 2006 already been experimenting with new online tools for a few years:

> The roots, if not the birth, of citizen-campaigning can be traced to the arrival of Howard Dean on the political landscape of the US in 2004. Despite ultimately failing to gain his party's nomination, his rise from unknown governor of a small north-eastern state to front-runner status in the Democratic primaries in late 2003 marked for many a "coming of age" of the Internet as a political medium. (Gibson, 2009:291)

Dean inspired Barack Obama, who went on to become the "social-media president" (Katz, Barris, & Jain, 2013). While, to the best of our knowledge, there have been no large-scale statistical analyses of the electoral effect of social media in the United States, it has been widely accepted by pundits and scholars alike that Obama at least in part won the presidential election because of his (team's) effective use of social media. As Gibson and McAllister (2011:227) note, "[t]he highly sophisticated online community-building and fundraising efforts of Barack Obama in the 2008 US presidential election are widely seen as having set a new standard for e-campaigning and have led to a scramble among politicians elsewhere to try to emulate his success." Our interviews with Dutch social-media managers confirm that the United States set the standard: virtually all of them mentioned Obama as their main inspiration.[19] For instance, the social-media manager of the Christian center right CU said: "I graduated in 2008, so it made me very enthusiastic. I think our whole generation, that is people who work with social media and politics, will remain so. I even wrote my thesis about his campaign. It was all so impressive." (CU, social media manager, 2014) His colleague from the postmaterialist Party for the Animals (PvdD) was also in awe: "This may sound a bit boring, but I read the book about the social-media campaign of Obama's first election from front to back!" (PvdD, social media manager, 2014). Even though there is little statistical evidence of the electoral effect of social media in American campaigns, the social-media managers we interviewed do believe that social media matter. Our analyses and the comparative evidence presented in this section both suggest they may be right—at least from the candidates' perspective.[20]

Conclusion

While most pundits are interested in the effects of social media on party vote shares (interparty inequalities), the effects of personalized communication tools such as social media may also have significant consequences for individual candidates (intraparty inequalities). Carrying out statistical tests of the interparty effects of social media is a cumbersome endeavor, one that is plagued by a "small N"-problem; it is more feasible to assess the intraparty effect. This is what we have done in this chapter, showing that on average social-media use does convince some voters, which suggests that more professional use has a more considerable impact.

In line with our presentation of the different (e.g., direct and indirect) effects of social media, our findings suggest that different mechanisms are at play at different times for different types of candidates. For instance, the direct-advertisement effect seems to have played a role for candidates in 2010, but less so in 2012, and an indirect effect must also have played a role in 2010 given the overall effect size compared to the number of followers (see Spierings & Jacobs, 2014). More detailed analyses with data on tweets being covered in traditional media and with voter-level data are useful in this respect. The analyses also showed that in the early-adoption phase it was mainly "naturals" and popular candidates who with some certainty benefited from using Twitter. By 2012, the parties and candidates had professionalized significantly, and more voters were active on social media. The electoral effect became more modest as competition on social media became fiercer and candidates no longer got a modernity bonus simply by being present. At the same time, the effect of general Twitter use became more widespread, which is not surprising given the larger public signing up to Twitter.

The latter finding goes against the overall equalization-then-normalization logic outlined by studies focusing on the interparty competition (cf. chapter 2). Yet it makes perfect sense that interparty normalization triggers intraparty equalization. Indeed, once major parties see the benefit of investing in social media, they start devoting resources and providing professional support to their candidates. This particularly empowers the lesser-known candidates compared to the naturals. We have shown that candidates with a high number of followers mostly benefited from tweeting in 2010, while in 2012 candidates with smaller networks were also able to translate their networks into more votes by tweeting actively. Interestingly enough, our analyses also suggest that underrepresented groups such as women actually benefited *more* than their overrepresented counterparts. This can probably be explained by the gender (in)balance within parties and traditional media, increasing the added value of social media for women as a communication tool that allows them to connect to their potential electorate.

All in all, based on these results the reluctance of more centralized populist parties, such as the socialist SP and the populist radical right PVV, seems "justified" (see chapters 4 and 5). After all, allowing individual candidates to tweet professionally also allows them to build a personal following and thereby to become a potential threat to the party leadership.[21]

Are these findings unique to the Netherlands? Given the importance of the electoral effects of social media, it is quite surprising that so few studies have been devoted to the topic. We identified 14 large-scale statistical studies of online campaigning and their findings are overwhelmingly consistent with our results, hinting at a modest positive effect.[22] However, more research is needed. For instance, we were the first to examine gendered effects, and it is thus unclear whether our findings are the exception or the rule. Furthermore, our diachronic comparison suggests that different mechanisms become more dominant depending on the phase of social-media diffusion in politics (and among the public). Detailed analyses are needed to test the ideas we generated from our analyses. What is beyond a doubt is that future research on preference voting should include controls for cyber campaigning on social media.

With an eye to the accumulation of knowledge, it would be useful to agree on a standard operationalization of social-media use. Our analyses suggest that is crucial to combine actual social-media behavior and content with the actual numbers of followers to avoid biases stemming from differences in offline popularity.[23] Experimental research is also needed to complement such research and examine which (if any) of the four mechanisms is at play. This will not be an easy thing to do, but some researchers are already embarking on the task (e.g., Kruikemeier et al., 2013). Last but not least, as most studies examine the effect of Twitter, it is useful to examine whether other social media have similar effects. Differences between different countries may pop up; in the Netherlands politicians, opinion leaders, and journalists are very keen on using Twitter, for instance. Studies outside the Netherlands have also found effects for Twitter (Jacobs et al., 2015; Suiter, 2015), but it may well be that the effect of other social media is "even" bigger in these countries, as high-quality Facebook use seems on the rise also in the Netherlands, particularly in terms of target-group strategies (see chapters 3 and 5).

CHAPTER 8

Conclusion

Introduction

On April 12, 2015, Hillary Clinton announced her bid to become the next American president via Twitter. It was a simple and concise way of conveying such an important message, and it showed how much politicians' preferences regarding communication channels have changed. In the recent past, an exclusive television interview or a message on the politicians' website would be the preferred ways to make such a big announcement. Not so anymore. Social media have become entrenched in many people's lives and seem to have especially influenced politics and traditional media (Peterson, 2012). In this book we examined whether social media have transformed politics beyond the rather exceptional moment of announcing a presidential bid, defying all odds and winning the American presidential election (i.e., Obama 2008), or starting a revolution (Arab Spring). Instead, we focused on normal politics and asked the question: *How have social media transformed normal politics in Western democracies?* In these conclusions we will return to this question in several steps. In the first section below, we will summarize our key starting points, namely the characteristics of social media and the opportunities they offer, and link them to our theoretical framework: the motivation-resource-based diffusion model. In the two subsequent sections we will assess whether the theoretical expectations derived from the model were corroborated and we will provide answers to our three more explorative subquestions. In the final two sections we will examine whether social media have transformed normal politics, and we will provide suggestions for future research.

A Motivation-Resource-Based Diffusion Model

What Makes Social Media Unique?

Throughout the book, we have used a simple definition of social media: Online platforms that allow a user to send, share, and consume information.[1] As individual social-media platforms often see a rapid rise and sometimes an even more rapid fall, we focused on general characteristics that distinguish social media as a broad category. We specifically highlighted five core characteristics of social media, two of which they share with traditional online media such as personal websites and mailing lists. Indeed, both Web 1.0 and 2.0 tools are *unmediated* and *personal*. But social media have more to offer, as some of their features clearly move beyond traditional online tools. For instance, where Web 1.0 also allowed for *interactivity*, this was always an option rather than the default setting. Social media center on interactivity and engagement, whether this is in the form of likes, shares, retweets, comments, or simple replies. Similarly, while email and personal websites were cheaper and easier to use than large-scale television advertisement campaigns, it still required a lot of expertise and money either to build professional websites or to collect large databases of email addresses (Gibson & McAllister, 2014). Compared to these "older" communication tools, social media are typically *cheap* and *easy to use*. In addition to these features, social media have something that other media hardly have: the virality potential. The speed and volume of social media, and their ability to snowball from first-degree friends to friends of these friends, all help reach people that normally would not have been reached.

Four Opportunities

These five characteristics offer new opportunities for individual politicians and parties alike. The degree to which these opportunities are fulfilled greatly affects the transformative power of social media. (1) In their simplest form, social media offer an *advertisement* opportunity, whereby social media are merely an addition to the communication toolbox. They help to communicate with people directly, but also showcase the politician or party as a modern actor who knows (and wants to know) what is going on in society. (2) Social media are renowned for their interactive opportunities and can simulate *human contact*. Such contact is still more indirect than face-to-face contact, but it is clearly less impersonal than watching a politician on television. Social media thus allow politicians to build a "social presence" (cf. Kruikemeier, 2014). (3) Instead of interacting with citizens in general, politicians and parties can also carve out a niche and *target* specific *groups* in society. This opportunity is especially useful when these groups are geographically scattered and difficult to reach via traditional media, such as youngsters, self-employed people, ethnic minorities, or LGBT people. (4) A specific target group that politicians are very eager to reach out to consists of journalists and opinion leaders. Indeed, social media are often populated by "movers" and "shakers," people who shape the agenda of the traditional media (PvdA, social media manager, 2013). In that sense, social media offer the opportunity to enter the *salon debate*, to use the analogy of the French salons in the eighteenth century, and thus influence public discourse.

The Motivation-Resource-Based Diffusion Model

The extent to which these four opportunities are utilized by political actors is crucial in understanding who favors social media and whether or not social media level the playing field. So far, academic studies analyzed specifically whether social media affect the interparty power balance. Some scholars have argued that technological innovations—particularly social media because of their low costs and easy access—favor previously disadvantaged players (equalization), while others claim that existing inequalities are simply mirrored on social media (cf. Gibson & McAllister, 2014; Schweitzer, 2011; Small, 2008; Vergeer & Hermans, 2013). Starting from Gibson and McAllister (2014), we built a multistep model of social-media diffusion which centers on the actors' motivation and resources, and helps to understand why and when equalization and normalization take place.

We distinguished three diffusion phases: the early-adoption, the widespread-use, and the laggard phases. In each of these three phases we expected different dynamics to be at play based on the motivation and the resources of the actors. (1) In the early-adoption phase, highly motivated and tech-savvy postmaterialist parties jump ahead of the other parties (leading to equalization for some). The bigger and non-postmaterialist smaller parties were expected to lag behind because they do not have the motivation to open social-media accounts. This motivation can be intrinsic (e.g., because the characteristics of social media fit postmaterialist ideology and principles) or extrinsic (e.g., because the postmaterialists' electorate is already on social media).[2] (2) In the widespread-use phase, when more people have social-media accounts, we argue that bigger parties suddenly have an electoral incentive (cf. extrinsic motivation) to join social media. Moreover, they have the money, personnel, and expertise to catch up with the postmaterialists and thereby become part of the leading group. The non-postmaterialist smaller parties on the other hand are not able to catch up and largely remain offline (leading to normalization for some, "unequalization" for others).[3] (3) In the last phase, the laggards finally start to adopt social media in larger numbers because *not* using social media would be considered old-fashioned. However, because it is late in the game and they lack both the funding of the bigger parties and the tech savvy of the postmaterialists, their social-media use is of relatively low quality. In this phase, some smaller parties (the postmaterialists) are still ahead of the other smaller parties (the others), but the bigger parties are on par with the postmaterialists (leading to equalization for some, normalization for most).

Theory Testing: Meeting the Expectations?

In order to answer our research question, we examined five interparty expectations, which were derived from our motivation-resource-based diffusion model. Table 8.1 summarizes and evaluates our findings based on chapters 4 and 6.

Early Adoption: The Postmaterialists Soar

In the early-adoption phase we expected to see bigger parties lagging behind while some smaller parties—particularly the postmaterialist ones—take the lead. At the *national level*, we found that the Greens were indeed well ahead of the competition. The other postmaterialist parties were on par with the bigger parties, all of which were clearly behind the Greens. Indeed the bigger parties adopted a "wait and see" approach, or as one of their social-media managers put it:

> *We are no early adopters and that is a deliberate choice. We go where our voters are. Once a network starts to take off, we'll go there.* (VVD, social media manager, 2013)

In that sense there clearly was some degree of equalization, but mainly for the Greens.[4] At the *local level*, all parties have less motivation to use social media, as their geographically smaller setting also gives them the option to contact citizens directly: The overall social-media adoption rates are clearly lower than at the national and European level. That

Table 8.1   Evaluation of the expectations

	National level	Local level	European level
Exp. (1) Early-adoption phase Equalization takes place, the bigger parties do not have the motivation.	+	+	+
Exp. (2) Early-adoption phase Postmaterialists do exceptionally well.	+	+	++
Exp. (3) Widespread-diffusion phase Normalization, bigger parties now have the motivation and the resources to buy expertise.	++	+/–[a]	++
Exp. (4) Widespread-diffusion phase Other smaller parties will not be able to catch up with the postmaterialists and bigger parties.	–[b]	+	++
Exp. (5) Laggard-diffusion phase The most traditional small parties will also start using social media, but lack of resources and expertise, and will remain behind.	++	n.a.[c]	++

[a] The biggest parties at the local level are often local parties. Yet they do not have extensive financial resources at their disposal. Small local branches of big national parties benefited from links with their mother parties.
[b] One non-postmaterialist smaller party, the ChristenUnie (CU), managed to catch up as well.
[c] This phase has not yet been reached at the local level.
Note: ++ = corroborated; + = mostly corroborated; +/– = evidence points both directions; — = mostly not corroborated;—not corroborated.

being said, all of the postmaterialist parties clearly stand out and the bigger parties lagged behind. At the *European level*, we found the most pronounced patterns: the postmaterialist parties were disproportionally active, while the bigger and other parties were largely absent or clearly lagged behind. Again there were clear signs of equalization, but not for all of the smaller parties. On the one hand, Expectations 1 and 2 are therefore by and large corroborated by our analyses: the postmaterialists are ahead, while the bigger parties do not have the motivation to catch up. On the other hand, these analyses clearly serve as a warning sign to those who want to lump together all of the smaller parties: the motivation and tech savvy of the postmaterialists are crucial factors in this phase.

Widespread Use: The Bigger Parties Catch Up, But...

Moving on to the widespread-use phase, the *national-level* analyses clearly showed that the bigger parties at this point started to invest in social media. Our interviews corroborated these data: the social-media managers of the three bigger parties all indicated that their parties started to invest in social media from 2011 onward, when social media started taking off in the Netherlands. However, the bigger parties were not the only ones who were able to catch up. Not surprisingly the two remaining postmaterialist parties rose to the top, but the small Christian CU party also did surprisingly well. Our interview with their tech-savvy social-media manager indicated that the party benefited from the support of a large kindred organization that had the expertise to help them out. The case highlights that social media are indeed cheaper than other media—if this were not the case, it would have been impossible to compensate for the lack of resources—but that some form of compensation is still needed. It also highlights the role of intrinsic motivation and the role of individual actors: the social-media manager of the CU started as a volunteer, convinced the head of the communication department, and took the party along with him.

At the *local level*, it is far harder for one person to make a difference and steer all the local candidates.[5] Not surprisingly then, at that level the pattern for the (non-postmaterialist) smaller parties is more in line with our expectations: They remained firmly behind the leading group. In fact, even the bigger ones had difficulties catching up. As budgets are smaller and campaigners are more often volunteers than professionals, even the local branches of the three bigger parties at the national level often struggled, though most of them managed in the end. The independent local parties, often the biggest parties in their municipality, fared worse: they did not have a national fall-back option and had to figure everything out for themselves. As a result, few of them managed to catch up with the postmaterialists. At the *European level*, the patterns are (again) the most straightforward: the bigger parties started to invest in social media and caught up, while the smaller non-postmaterialist parties came nowhere near the other two groups of parties. Overall, expectation 3 is mostly corroborated, while the exception of the CU shows that smaller parties can catch up (contrary to expectation 5) if they find a substitute for the resources needed to catch up.

Laggard Phase: Reluctantly They Follow

Lastly, in the laggard phase we expected the postmaterialist and bigger parties to start improving the quality of their social-media use by better utilizing the opportunities social media offer. At the *national level*, this expectation holds very well. The parties who were ahead started to develop more complex and high-quality strategies, but the smaller inactive parties struggled. Both the smaller party for the elderly (50Plus) and the ultraorthodox Christian party SGP opened Twitter and Facebook accounts but only made haphazard use of them.[6] Similarly, the populist parties' politicians were active on social media, but did not make use of its more advanced opportunities. One of them, the PVV, seemed to consider Twitter merely as an alternative to sending out press releases. As we said earlier, such selective use of social media makes it nearly impossible to reap the benefits of its unique opportunities. At the *European level*, we observed similar trends: by 2015 the laggards mostly had social-media accounts, or at least a Twitter account, but the quality of their social-media use was lower. While similar to the national level, this trend is not identical: the advantage for the postmaterialists and the disadvantage for the populists seem to be more pronounced. Furthermore, among the bigger parties the Christian democrats seem to do particularly well. The latter is not that surprising as they are part of the largest European party (EPP) and received their support (CDA, social media manager, 2014).[7] Again, receiving a helping hand from someone can make a big difference. Overall then, Expectation 5 is corroborated: Those that fall behind will remain behind.

Theory Building: Moving Beyond National Interparty Relations

While an important part of this book dealt with refining and testing the classic interparty equalization versus normalization framework, an equally important part centered on questions that went beyond this classic framework and were aimed at theory building. In this vein, we started from our motivation-resource-based diffusion model to ask three additional research subquestions in chapter 2. Specifically, we formulated subquestions dealing with the impact of the nature of the specific social medium (e.g., Twitter versus Facebook), the nature of the arena one examines (e.g., local versus national), and the nature of the unit of analysis one examines (e.g., politician versus party). In what follows we address these three subquestions in reverse order—from specific to broad—as the answers to the more specific subquestion(s) allow us to address the broader one(s).

Before we proceed we would like to offer a word of caution. As we are dealing with theory building in this section, our findings are more explorative by definition. Sometimes the data available had severe limitations, sometimes the number of cases was relatively small, and overall they only pertain to the Netherlands. Nevertheless, the findings often point to a clear pattern and, given that the Netherlands is a typical case regarding many important background characteristics, we would not be surprised if scholars examining other cases will find the same patterns. As such, the findings should be considered a starting point rather than a terminus.

Personalization, Equalization, and Normalization in IntraParty Politics

As outlined in chapter 5, there often is a tension between *intraparty empirics* and *interparty theories* in research on social media and politics. Indeed, most studies examining the diffusion of social media in politics adopt an implicit intraparty perspective. They analyze individual politicians' adoption and usage rates, while the role of parties is often left out of the picture. Yet typically such studies apply an interparty theory to interpret their findings: The equalization versus normalization framework. To ease this tension we suggested two solutions: first to add a perspective that explicitly deals with the relationship between parties and politicians (the personalization perspective) and second by systematically applying the logic of the equalization-normalization logic to the intraparty level. This opens a range of new questions, such as: who benefits most from social media, top candidates or the more disadvantaged ones; candidates from traditionally overrepresented sociocultural groups or the more disadvantaged ones?

The personalization literature suggests a shift from party to politicians, which cannot but create a certain tension, as parties want to control the central party message (cf. Norris et al., 1999). Our interviews with social-media managers indicated that parties use three strategies to ease this tension: (1) they can limit the number of politicians on social media by forbidding and discouraging them, (2) they can monitor and control them, and (3) they can train them. Particularly *centralized parties* (such as populist parties) opt for the first strategy. Surprisingly, all of the parties control or monitor their politicians in some sense, while parties who catch up in the widespread-diffusion phase mainly use the third strategy, probably because the candidates of such parties lack the intrinsic motivation and are less tech savvy than the early adopters. The postmaterialist parties seem least worried about losing control.

Moving to our intraparty findings, Table 8.2 presents an overview of our national-level (widespread-diffusion phase) findings from chapters 5 and 7.[8] The first three rows of Table 8.2 summarize our findings on

Table 8.2 The intrapartyd imension

	Network		Use	Electoral benefit	
	Tw	*FB*	*Tw*	*2010*	*2012*
List pullers in the lead from widespread use onwards (*concentrated personalization*)	++	+/–[c]	++[d]	++[f]	++
Top ten candidates catch up (*diffuse personalization*)	+[a]	+/–	+	+/–	+
Lower-ranked candidates catch up (*diffuse personalization*)	+/–[a]	–	+	–	+
Female candidates do better than men (*representational equalization*)	+/–[b]	n.a.	+	+/–	++
Ethnic-minorities candidates do better than ethnic-majority candidates (*representational equalization*)	+	+	+/–[e]	n.a.	n.a.

[a] Some mid-tier or even lower-ranked candidates were able to collect a substantive Twitter network, though especially most of the lower-ranked ones faced an uphill struggle.
[b] Women are more likely to have a Twitter account, but men have more followers.
[c] Facebook is mostly used by the parties, not by candidates, and the number of likes/friends is markedly lower than the number of Twitter followers.
[d] The Dutch candidates all tweeted a lot, and while list pullers tweeted the most, even the lower-ranked candidates sent out a lot of tweets.
[e] Holds for ethnic-minorities women, not for ethnic-minority men.
[f] in 2010 the number of followers was crucial (which benefits the list pullers), in 2012 the effect of tweeting was more important (which benefits everybody).

Note 1: ++ = corroborated; + = mostly corroborated; +/– = evidence points both directions;—= mostly not corroborated;—not corroborated.
Note 2: "List puller" = position 1; "Top ten" = position 2 to 10; "lower-ranked" = remainder.

which candidates use and benefit the most from social media. We find clear indications of a *strong concentrated personalization* (cf. normalization) and *some diffuse personalization* (cf. equalization). The usage patterns are very clear on Twitter: the list pullers use the medium the most and have by far the most followers. However, the other top-ten candidates were also able to build a substantial network through active usage. Even some of the lower-ranked politicians managed to do so, though it has to be noted that it was more difficult for them and took them more time and required higher-quality usage. Interestingly, Facebook is hardly used by the candidates and is mainly a tool for the parties. While list pullers do have Facebook pages, the party headquarters often maintain the page, and the number of likes it gets is only a pittance compared to the vast droves that follow them on Twitter.

We also looked at the electoral benefit of social-media use. Here we found that in 2010 the candidates with large numbers of followers mainly managed to do well. Some lower-ranked and mid-tier politicians managed to build a reasonably sized network, but the list pullers benefited most. By 2012, tweeting became more important and parties started to support their candidates across the board, which somewhat reduced the differences between the list pullers and the other candidates. The individual candidates now were more in control of their fates. In sum, we find normalization followed by (some) equalization. In chapter 7, we suggested that the mechanism explaining this intriguing finding is as follows: because parties start to see the added value of social media, they start to invest in *supporting their candidates*. Hence increasing interparty normalization, whereby all the parties are on social media, entails a certain degree of intraparty equalization.

Table 8.2 summarizes our findings regarding social-media use by underrepresented groups. We specifically analyzed women and ethnic-minority candidates. On the one hand, women struggle to build large networks, they use social media more often, *and*, electorally speaking, benefit on average almost twice as much from tweeting than their male counterparts. Ethnic-minorities candidates, on the other hand, had relatively large networks on Twitter and Facebook, but did not use these accounts more than their ethnic-majority counterparts. Unfortunately given the low number of ethnic-minorities candidates we could not analyze whether or not they benefit more electorally, but this is definitely an important venue for future research.

We can now answer the most specific of our research sub-questions (RSQ3): *How do social media change the power balance of individual politicians within parties vis-à-vis each other and the party organization?* Overall, most power still remains in the hands of the party leadership and the list pullers: they have developed strategies to counter the risk of losing control and are able to attract most of the attention on social media. Yet several previously disadvantaged candidates managed to chip away some of the party headquarters' power. In that sense, social media did trigger at least some degree of intraparty equalization.

First- and Second-order Political Arenas

So far we have discussed the national, first-order political arena, but what about other political arenas? This book was the first to systematically assess and compare interparty dynamics in multiple arenas (see chapter 6 and Table 8.1). The trends at the local level are similar to those at the national one: Concentrated personalization is the rule, diffuse personalization the exception. However, the difference is less pronounced. Some lower-ranked candidates manage to catch up easily. Furthermore, being an actively tweeting candidate is even more important for building a larger network at the local level, which suggests that politicians can take matters into their own hands. More generally, fewer candidates are active on social media here, probably because face-to-face contact is a viable alternative and the degree of professionalism in politics is lower.

Regarding the European level, face-to-face contact is not an option. Not surprisingly then, by 2015 the MEPs were frantically making use of social media, Twitter in particular. Again, some degree of concentrated personalization was taking place (especially for the laggard parties SGP, PVV, and SP) though the other candidates managed to draw a lot of attention as well. Interestingly, chapter 6 showed that the playing field is actually quite leveled and that the female MEPs did especially well: they had high numbers of followers, tweeted often, and triggered a lot of engagement from their networks. Our interviews indicated that as the relatively small pool of MEPs get support from both the European parties and the national headquarters, they thus have a fair amount of financial resources and expertise at their disposal. They also have a high extrinsic motivation: traditional media ignore European politicians and face-to-face contact is not an option. As such it should not come as a surprise that the MEPs are among the most social-media-savvy politicians.

Combining this information with the evaluation in Table 8.1 in the previous section we can answer our second research subquestion (RSQ2): *How does political context affect the impact of social media on the power-balance position of parties and politicians?* One of our most important findings is that we *should not lump together* the different second-order arenas. Regarding interparty politics, the local arena is slower to adapt to social media because it is more geographically diverse, involves more voluntary rather than professional politicians, and because face-to-face contact is an option. The European arena involves fewer politicians, is more professionalized, and does not have the option to revert to personal contact. However, the overall interparty patterns point in the same direction as our national analyses, an regarding intraparty politics the same conclusion holds: Concentrated personalization is the rule, diffuse personalization the exception. That being said, our European analyses lead us to cautiously support the claim that equalization is possible, especially for female politicians. Lastly, it should again be stressed that there clearly was less data available for the second-order political arenas. In short, three lessons *can* be drawn: (1) second-order arenas should not be lumped together; (2) dynamics in second-order arenas can only be understood knowing the first-order political inequalities; (3) patterns point in the same direction, but notable differences appear.

It is important to carry out follow-up research as our explorative analyses revealed some intriguing trends.

Different Social Media, Different Patterns?

In chapter 2, we noted that different social media may well have different effects. In this book we focused on the two most important social media for politicians: Twitter and Facebook. Whereas Facebook is a complex personal pegboard, Twitter is more of a personal press agency. Both have distinct advantages and as such we expected that parties and politicians would use them differently. To a certain extent this was the case, but the most important observation was that Facebook was simply used less and was considered to be a private communication tool, certainly in the early-adoption phase, whereas Twitter was considered to be a professional communication tool. Moreover, using Facebook professionally is more complex and requires more expertise. It should not come as a surprise then that *parties* used Facebook quite often, as they were particularly charmed by its ability to reach a lot of people (and not just opinion leaders). Still, some candidates did make good use of Facebook, such as some of the ethnic-minority candidates. Moreover, given that Facebook is more difficult to use, politicians are more dependent on the parties to provide them with professional content (e.g., pictures, infographics, etc.). Twitter also has one last major advantage over Facebook in the Netherlands: journalists are always on Twitter, but rarely on Facebook (cf. salon-debate opportunity).

Based on this summary we can answer the last, more overarching research subquestion (RSQ1): *Is there a connection between the different functionalities of social-media platforms, most particularly Facebook and Twitter, and the degree to which they lead to equalization or normalization in the different phases of diffusion?* There clearly is a connection between the functionalities of social media and the ways the parties and the politicians use them, but it is not just the functionality that counts, the investment a politician needs to make is also crucial. In fact, the connection is so strong that relatively few Dutch politicians use Facebook politically. Facebook requires more advanced use, especially in the later diffusion phases. In the early-adoption phase, Facebook itself was still growing in the Netherlands, as a local variant (Hyves) was still much more popular. By 2012, both the bigger and the postmaterialist *parties* discovered the medium, but their politicians only used it privately. One of the postmaterialist parties, the progressive liberal D66, even made this its official party policy. At the same time the bigger parties were devoting far more money to the medium and outspent their opponents. Interestingly, even in the laggard phase, the inactive smaller parties hardly used Facebook. Twitter, however, was used by both candidates and journalists to engage with each other, and its diffusion was quicker. All in all then Facebook is a social medium that tilts more toward normalization, whereas Twitter offers more opportunities for equalization. A wider study of these differences and their effects in terms of building long-term party identification and short-term vote gains is needed. For instance, the exceptional usage of Facebook by some candidates does raise the question whether they benefit more than others from social media, as only a few politicians are on Facebook (cf. Twitter's impact on votes in 2010 [chapter 7]).

Have Social Media Transformed Politics?

Let us now return to our overall research question: *How have social media transformed normal politics in Western democracies?* Did they reduce political inequalities within and between parties? While to a large degree much remained the same, social media undeniably did have some effects. In this section we address the five main areas where social media had an impact.

1. Interparty Politics: Populists as the Losers of Digitalization
As expected by scholars such as Gibson and McAllister (2014), the Dutch Greens performed extraordinary well on social media. More generally, postmaterialists do well on social media, regardless of whether they are very small (Party for the Animals, PvdD) or relatively big (the progressive liberals, D66). The bigger parties will not be left behind and will be able to buy their way in, but some parties are left behind. This to some extent holds for the smaller non-postmaterialist parties, but a fortiori for the populist parties. This has very little to do with their ideology but more with their party structure, as those parties are the most centralized. They are caught between a rock and a hard place: allowing their candidates to use social media professionally would potentially undermine the party leadership, while not doing so means lagging behind their competitors.

2. Intraparty politics: Follow the leader . . .
At the national level, the most popular politicians on Twitter are the list pullers: most citizens seem to follow the party leaders. However, some politicians managed to carve out their own niche and build their own electorate. In that sense, social media offer politicians the opportunity

to build their own electorate. Even if they do not win enough preference votes to get elected directly, it may increase their intraparty position and signal to the party leadership that they are a strong candidate. Hence, while it is tempting to zoom in on the dominance of list pullers on social media, below the surface some things seem to be changing, and these other candidates that are performing well deserve more attention.

3. Intraparty Politics: It's a Woman's World?

This conclusion—it's a woman's world—seems to hold especially for female politicians in the Netherlands. While male candidates had more followers, female candidates tweeted more and yielded more electoral benefits from social-media use. At the same time, patriarchy will be not overthrown directly; gender equalization is relative and the list pullers are generally men. We were unable to examine whether other underrepresented groups also benefited more, but if this pattern surfaces in further research, this is quite an intriguing effect that may reduce offline inequalities.

4. Intraparty Politics: A Handful of Votes

One of the questions we often get asked by pundits, journalists, and politicians is whether social-media use yields an electoral bonus. As chapter 7 illustrates, there surely seems to be a modest (but statistically significant and therefore consistent) correlation between Twitter activity and the share of preference votes a candidate gets. Moreover, by 2012 tweeting had become more important than the number of followers a candidate had, which offers some opportunities for lower-ranked candidates to change their fates. In sum, social media did not transform electoral campaigning, but they did add a useful tool to the individual politicians' communication toolbox.

5. Second-order Arenas: Local Transformation: No; European Transformation: Yes

The transformative power of social media was most pronounced at the European level: list pullers felt the lower-ranked candidates breathing down their necks, and female MEPs also did well. At the local level, the impact was less visible, and the only parties that seemed to have mastered social media were always the postmaterialist parties.

Comparative Perspective: The Netherlands as a Harbinger of Things to Come

Do our findings travel to other Western democracies? In each of the chapters we included a comparative section, and these show that our framework not only fits the Dutch case but that the bits and pieces available on other polities fit the framework too. This suggests that the broader patterns hold across countries. Other Western democracies also saw postmaterialists and bigger parties do well, saw populists perform badly, witnessed more concentrated than diffuse personalization, found a correlation between social-media use and votes, and noticed that local politics adapts more slowly to social media, while European politics actually adapts them faster. The main difference between the Netherlands and most other Western democracies, however, is that Dutch society is ahead of many other countries when it comes to social-media use (cf. chapter 3), which might mainly imply that what is happening in the Netherlands is likely to happen in other countries as well.

The American Exception: Obama, Inspiring but Unlikely

One country was clearly ahead of the Netherlands: the US. Social media are widely used in the United States, and political consultants and campaigners look to the United States for inspiration. Indeed, virtually all of the social-media managers we interviewed cited the Obama campaign as their main source of inspiration. Yet there is a certain irony here. In our comparative sections, we noted that the political culture and system of the United States actually make the country an outlier on many topics: smaller parties stand no chance, minorities are way more active, and politicians used social media in an unusually personal way. Moreover, the successful Obama campaign—typically seen as the best example of successful social-media use—may not be easy to apply elsewhere. While inspirational, this campaign required advanced technical tools and lots of staff, and thus was far from cheap. In countries that do not have liberal campaign-financing legislation or where the number of campaign volunteers is smaller, the lessons from the Obama campaign are hardly applicable.

The Road Ahead

This book offered a rich in-depth quantitative and qualitative case study of the Netherlands in comparative perspective. Along the road many questions popped up, triggering follow-up questions. Some of these we could answer; for others we lacked the data. Some of our results are strong and seem corroborated by other research, as indicated throughout the book. Other times, our research was more explorative and research on other countries was largely absent. We would thus like to draw attention to the following theoretical, methodological, and empirical avenues for further research.

Theoretical Avenues

In our more explorative sections, several possibilities for future theoretical development have already popped up. In this brief section we reiterate three of the most prominent ones. First, it is useful to examine the impact of different social media further. Social media are often lumped together but the specifications of the each medium may affect its impact. While we have presented a brief exploration of the topic, more can and should be done on the matter. For instance, though we found that most candidates do not use Facebook, some do. Why is this the case and what is the impact of such use? Second, though the poor showing of populists is intriguing, there are differences within the group of populists. Are these differences a matter of party centralization or do they also have genuine ideological foundations, for instance due to the difference between left versus right populists? After all, the latter lean stronger toward authority and order. Or maybe the type of party leader plays a role? Third, we have observed some notable gender differences, but the mechanism why women win more by using social media is still largely uncharted territory. Is it because they tend to be more interactive? Are male politicians more "self-obsessed"? Might women revert to social media because they have less access to traditional media? Or are the differences voter related? Maybe women are on average more present on social media?

Methodological Avenues

During our research for this book, we encountered some methodological complexities that deserve more attention in future research. First, most research nowadays focuses on Twitter. Does this focus on Twitter bias our findings? Should we focus more on Facebook? And if so, how to deal with personal accounts that are shielded, while organizational pages are public? Moreover, how can we judge what good Facebook positing behavior is if Facebook's algorithm is secret and constantly changing? Can it be that new types of social media are actually gaining traction? Are new messaging systems such as Facebook Messenger, Snapchat, or WhatsApp useful for politicians? And, methodologically, how to examine them? What is important here is that these messaging systems are a form of one-on-one communication. This raises not only technical but also ethical questions about whether researchers should start following or connecting with the people they study? Second, how to conceptualize, operationalize, and measure normalization and equalization? The normalization-equalization debate is mostly understood in binary terms: either normalization occurs or equalization. However, as we have seen, it matters greatly which parties one compares: some smaller parties do well compared to others, while they may still only be on par with the bigger parties. Is this normalization or equalization? If a local party is the biggest in a municipality but does badly compared to a moderately strong local branch of a national party, is that equalization or normalization? If a smaller party is on par with a bigger party, is that equalization or normalization? After all, in the offline world, the bigger party is bigger than the smaller one, which means being on par is actually quite an accomplishment already. As we have shown, things can get messy in multiparty systems, and more research examining such cases is therefore important, as is conceptual refinement. Third, this book used in-depth comparison based on a wealth of information about candidates and parties. While this is possible in a book-sized study, journal articles generally do not have enough space to explain all this. This becomes even more problematic when different data and measurements are available across countries. In other words, one of the core challenges to the field is to develop more standardized operationalizations of quantitative and qualitative diffusion.

Empirical Avenues

Lastly, our research suggested some starting points for future empirical research. First and most obviously, our motivation-resource-based diffusion model needs to be tested in other settings. Do the same results pop up elsewhere? Do the different diffusion phases also witness such distinct patterns in other countries? Second, the theoretical interpretations in our sections that were geared toward theory building need to be corroborated in other settings. Typically we had to work with limited data, which urges other research examining the same questions. Third, it seemed that the Netherlands is a ferociously Twitter-savvy country. If other countries are more Facebook-savvy, does that mean Facebook takes over the functions of Twitter (i.e., the salon debate-opportunity)? And how does this influence diffusion patterns?

As always, research calls for a lot of follow-up research. This does not mean we did not find any answers. Cyberspace, social media, and politics might all be rapidly changing, but we found that distinct patterns can be observed below the surface and similarities across countries and time occur more often than one would imagine. That we—academia—are still able to lay bare those patterns and help understand how technological innovations alter (some of) the political inequalities, is in itself a comforting thought.

Appendix:
In-depth Expert Interviews

Name	Party	National or local	Position	Date	Interviewer	In person or telephone
Mei Li Vos	PvdA	National	Active, high profile politica	06/07/2013	Kristof Jacobs	In-person
Pia Dijkstra	D66	National	Active, high profile politica	04/11/2013	Kristof Jacobs	Phone
Maarten Hijink	SP	National	Campaign leader for social media	25/11/2013	Kristof Jacobs	In-person
Jaap Stronks	Independent/ PvdA	National	Social media consultant; advisor PvdA	26/03/2014	Kristof Jacobs	In-person
Floris Spronk	CU	National	Campaign leader for social media	25/04/2014	Kristof Jacobs	In-person
Frank Magdelyns	Inwonerspartij Toekomst Houten	Local: Houten	Locally responsible for online campaign	25/07/2014	Jeroen Hellebrekers	Phone
Marike van der Leeden	Stadspartij Wageningen	Local: Wageningen	Locally responsible for social media and council candidate	04/08/2014	Jeroen Hellebrekers	Phone
Marjet Woldhuis	Stadspartij Groningen	Local: Groningen	Campaigner and council candidate	05/08/2014	Jeroen Hellebrekers	Phone
Sabine Koebrugge	VVD	Local: Groningen	Campaign leader and council candidate	08/08/2014	Jeroen Hellebrekers	Phone
Eef Stiekema	VVD	Local: Houten	Locally responsible for for social media and council candidate	08/08/2014	Jeroen Hellebrekers	Phone
Lodewijk Bleijerveld	PvdA	National	Campaign leader for social media	21/08/2014	Kristof Jacobs	In-person
Huub Bellemakers	GroenLinks	National	Campaign leader for social media	13/09/2014	Kristof Jacobs	In-person
Arjan Vliegenthart	SP	National	Campaign leader and national politician	24/09/2014	Kristof Jacobs	In-person
Arnaud Proos	SGP	National	Campaign leader for social media	25/09/2014	Kristof Jacobs	In-person
Tim Versnel	VVD	National	Campaign leader for social media	04/10/2014	Kristof Jacobs	In-person
Celina Kremer	CDA	National	Campaign leader for social media	14/10/2014	Kristof Jacobs	In-person
Eva van Esch	PvdD	National	Campaign leader for social media	21/11/2014	Kristof Jacobs	In-person
Dieder de Vries	D66	National	Social media webcare team	23/12/2014	Niels Spierings	In-person

Notes

1 Introduction

1. Howard and Parks (2012: 359) offer the following, more formal definition: "Social media consists of (a) the information infrastructure and tools used to produce and distribute content that has individual value but reflects shared values; (b) the content that takes the digital form of personal messages, news, ideas, that becomes cultural products; and (c) the people, organizations, and industries that produce and consume both the tools and the content."
2. This dominance of Anglo-Saxon cases also resonates in the terminology of the politics and social-media literature: most theories speak of large, small, and fringe parties. In multiparty systems, a substantial percentage of parties are actually *medium* in size (Farrell, 2011).
3. Interestingly enough, Dutch parties seem to mostly emphasize Facebook, whereas politicians prefer Twitter as a political medium. This is nicely illustrated by a Dutch campaign manager who said "[Facebook is] kind of the focal point. The party Twitter account is less important than the personal Twitter accounts of politicians." (PvdA, social media manager, 2013) This is supported by a prominent politician from that party: "Facebook is simply too complicated, Twitter is very easy to use... Twitter is one's personal press agency."
4. As the online campaign manager of the Dutch social democrats put it, talking about experimenting with Google Hangouts: "We just wanted to appear modern" (PvdA, social media manager, 2013).
5. For the Netherlands, the only significant example is a negative incident. In 2009, Arend-Jan Boekestijn, a prominent Dutch MP for the conservative liberals, tweeted in a discussion about dictators and the deaths for which Mao was responsible: "I sometimes overlook a slit-eye, there are so many of them!" Other social-media users, journalists, and politicians heavily criticized this racist remark and Boekestijn had to publically apologize. Together with two other incidents this contributed to his resignation later that year. While it illustrates the power of social media, the incident says little about the potential of social media to create large individual networks via social media.
6. http://www.internetworldstats.com/top25.htm (visited on March 17, 2015).
7. http://en.wikipedia.org/wiki/List_of_countries_by_number_of_broadband _Internet_subscriptions (visited on March 17, 2015).
8. For more information, see: www.ru.nl/VIRAL.

2 Theorizing Social Media, Parties, and Political Inequalities

1. Vergeer, Hermans, and Sams (2011:483) note that "although interaction was distinguished on traditional Web 1.0 websites, it merely facilitated interactivity in a passive way."
2. Admittedly not all services are free; for instance, obtaining statistics on one's outreach and effectiveness are hidden behind a paywall. It is also possible to invest money to increase you reach beyond the one the specific platform's algorithm grants you. For instance, an organization can "boost" a Facebook post by paying for it.
3. In that sense they are different from simple ground campaigning: Ground campaigning works best when the target audience is geographically concentrated.
4. The moderate skeptics do not believe in a positive effect but would not go as far as claiming there is a negative effect.
5. A *Twitter storm* is a sudden surge of attention for a tweet where a damaging message goes viral but quickly fades out.
6. We might actually distinguish between *social-media zombies* and *social-media vampires* here: social-media vampires are mostly dormant, and you won't see them when you are active on social media, whereas social media zombies are around but obviously not actually participation effectively in the social aspect of social media. Both types of politicians exist.
7. Early evidence on the United States actually suggests that social-media use is associated with *more* political interest and more political engagement (Gil de Zúñiga, Jung, & Valenzuela, 2012).
8. They are no hypotheses in the strict sense, as we are not just interested in associations but also in the mechanism and in the role of motivation and resource.
9. Starting from 5,000 friends, the system changes from two-sided befriending to one-sided liking.
10. Noteworthy exceptions are Larsson (2013), Mascheroni and Mattoni (2013), Suiter (2015), and Vergeer, Hermans, and Sams (2011).
11. As mentioned before, Facebook is more technical than Twitter and requires more resources. One can thus expect the resource advantage of top politicians to be more visible on Facebook than on Twitter.

3 Social Media in Politics: The Netherlands from a Comparative Perspective

1. For more detailed information about Dutch politics and the Dutch political system, we refer to existing works, such as Andeweg and Irwin (2005).
2. These numbers refer to the situation in 2015. The number of municipalities tends to fluctuate slightly due to municipal mergers.
3. The electoral quotas are calculated by dividing the total number of valid votes by the total number of seats available (e.g., 150 in the case of the Lower House).

4. Typically there are about 9.5 million voters who cast a valid vote.
5. www.kiesraad.nl.
6. IPU: Women in National Parliaments: http://www.ipu.org/wmn-e/arc/classif
   311010.htm and http://www.ipu.org/wmn-e/arc/classif010214.htm.
7. This was illustrated nicely by four low-ranked candidates whose campaign
   efforts were followed in 2012 by *De Volkskrant*, one of the main quality
   newspapers in the Netherlands. (De Vries; http://www.volkskrant.nl/dossier
   -vk-dossier-verkiezingen-van-2012/strijden-voor-een-zetel-jammer-dat-ik
   -mijn-campagne-cadillac-moet-inleveren~a3314515/).
8. A comment by the social democrat Mei Li Vos (PvdA) is telling in this
   respect: "My campaign will be mostly an online campaign." (De Vries,
   2012)." Another example is the progressive liberal politician Boris van der
   Ham (D66), who obtained 42,000 preference votes in 2010. In 2006 and
   2010 he was elected most active Internet politician (Dijkstra, 2012), and in
   2010, he waged an extensive online campaign, spreading personal videos via
   his personal website and social media. In both cases he was elected with a
   higher number of preference votes—though he was ranked high on the party
   list as well.
9. See for instance: Miniwatts Marketing Group (2015).
10. For instance, Vergeer, Hermans, and Sams (2011) compare the number of
    Twitter accounts in the Netherlands and the United States in 2010. The dif-
    ference is a factor of 20; exactly the difference in population size.
11. Measured on April 8, 2015. Only Google+ shows decent numbers, 20,000 fol-
    lowers for these two parties in early 2015.
12. See also: Weber Shandwick (2014:9).
13. One party, however, shielded its account for outsiders: The populist radical
    right PVV.
14. Again, the populist radical right party PVV did not have an official party
    account.
15. The two parties we were unable to interview seem to be the exception. The
    party for the elderly (50Plus) stopped tweeting after the 2012 elections, which
    might appear a bit opportunistic. The populist radical right party (PVV) has
    shielded its party account, which may be a way to hide their inactivity.
16. The shorter period biases the percentage upwards. If anything this bias further
    supports our conclusion.
17. For 33 of the 381 candidates for whom we found Facebook accounts, we could
    not retrieve information on their number of followers.
18. One solution in this respect is buying followers, at least that might have been
    what the CDA new list puller or his advisers thought when Van Haersma
    Buma became list puller and went from being rather unknown and only fol-
    lowed by roughly 4,500 accounts to almost 25,000 accounts in a matter of
    days, though 80 percent were robot accounts. His party denied having bought
    followers, but that is how it was covered on television, radio, newspapers, and
    the Internet (NOS, 2012).

19. In what follows we will understand professionalization as a use that (1) fits the medium (cf. causal characteristics, chapter 2), (2) adheres to overall quality criteria (e.g., sharp versus blurry pictures), and (3) maximizes the virality potential.

20. For this work, we specially thank our intern Eirin Vikki Kofoed. More info on the coding and the data are available via the authors.

21. A professional profile picture is defined as one that is in focus, well cropped, and professionally lit with a bright background.

22. This raises the question to what extent such an emphasis would actually benefit politicians. While no definitive answer can be given for the time being, Kruikemeier's 2014 analysis does suggest the electoral impact would be close to nonexistent.

23. These pictures show volunteers in party outfits in two cases, a party leader twice, and a picture symbolizing the party identity in the cases of the Greens and the Animal Party.

24. One party, the postmaterialist progressive liberal D66, even went as far as to institutionalize this by using Twitter question hours.

25. Own calculations based on Smith (2012).

26. Calculated as follows: comment+likes divided by number of posts and fans) (Smith, 2012).

27. See Smith (2012).

28. Admittedly even then posting many messages does not equal interaction. However from a 2009 study on the 36 Dutch EP candidates on Twitter we see that about 44 of every 100 messages included an at-mention (@), which is generally considered an indication of interactivity (Vergeer, Hermans, & Sams, 2011). While Kruikemeier (2014:136) found that in the 2010 election campaign only 28.5 percent of the messages used an @-mention, the number of tweets had risen substantially, so the absolute number of interactions was higher.

29. Jeroen van Wijngaarden (VVD Amsterdam, 2014). Tanja Jadnanasing is married to a man and does not identify herself as being bisexual.

30. Henk Leenders: Background picture includes banner "Together for Brabant"; Carola Schouten: profile text includes "Randstad Brabo" and "Rotterdammer with a Brabant heart"; Henk Nijboer's profile text includes "Groninger"; Sander de Rouwe's profile text includes "Frisian"; Agnes Mulders' profile text includes "The Hague could use a bit of Drenthe."

31. (1) Television: Jaap Jansen (ÉénVandaag); Frits Wester (RTL Nieuws); Jos Heymans (RTL Nieuws); Rick van de Westelaken (NOS Journaal); Vincent Rietbergen (NOS Journaal); Xander van der Wulp (NOS Journaal); (2) Radio: Lara Rense (Radio 1); Rachid Finge (NOS Radio); Derk Marseille (BNR Nieuwsradio); (3) Internet media: Jeroen Mirck (Joop.nl); Kaj Leers (Z24.nl); (4) Written Press: Bas Paternotte (HP/De Tijd) Bert Wagendorp (De Volkskrant); Kustaw Bessems (De Pers).

32. Due to data availability, we focus mostly on interactivity. Theoretically such a focus makes sense when studying Twitter (as we do here) (cf. Kruikemeier, 2014). Studying Facebook would require studying other elements as well.

33. Martin (2013) mentions 450 MPs being on Twitter. The House of Commons has 650 seats.
34. Given the lack of in-depth comparative studies or studies across countries, it is more difficult to draw strong conclusions about the professionalization or qualitative diffusion of social media. In terms of using the interactive features of social media, the differences between the Netherlands and the United States seem to favor the Netherlands as well, but in terms of personalization, the American candidates seem more prone to make use of the human-contact and target-group opportunities.

4 Interparty Relations: David versus Goliath

1. The candidate data are further explored in the next chapter, which tackles intraparty inequalities.
2. In a multiparty system, the labels "big," "small," and even "fringe" should be interpreted loosely, as the parties' vote shares are very close to each other.
3. Some also add the third category of "fringe" or even "non-parliamentary" parties, see for instance: Gibson, Römmele, and Williamson (2014), Schweitzer (2011), and Southern (2015). The theories used here, however, typically only make a two-way distinction.
4. This should not come as a surprise as the party is the smallest of the three and has been in steady decline.
5. As it was not always possible to provide such detailed data for more than one data point, it is quite difficult to draw conclusions about "diffusion" (which requires a longitudinal analysis).
6. Tellingly, the "three MPs have a Facebook page. It was a gift from our party's youth organization. (...) They told them half jokingly, "We'd like you to be more active on Facebook, so we already made a Facebook page for you" (SGP, social media manager, 2014).
7. It has to be stressed though that there is a discrepancy between the party during a campaign and a party after the campaign. In "peacetime," the team is much smaller and there is no professional webcare team, for instance (D66, social media webcare, 2014). Clearly it was a significant financial burden for the party to bear, one that was only bearable for a short period of time.
8. The social-media staff consisted of relatively cheap junior employees.
9. Table 4.4 also shows the limitations of the impact one person can have: In terms of media recognition and network size the CU is well behind of the bigger and postmaterialist parties.
10. So far Wilders seems to do well, as the PVV is the second party when it comes to journalists considering tweets newsworthy (cf. Table 4.3).
11. Remember that the smaller CU had a tech-savvy volunteer, but even he needed the support from the much larger Christian public broadcasting service.

5   Intraparty Relations: David versus Nabal and Abigail

1. Well-known politicians (president of parliament, chair of parliamentary research committee, spokesperson on topical debates, party leaders, party presidents, executive member of a local government or a major city, mayor); presidents of a highly visible NGO, labor union, or employer's organization; TV presenters or anchorpersons; and opinion leaders and prominent activists (see Spierings & Jacobs, 2014).
2. No self-identified transgender candidates (who identify neither as woman nor as man) were running.
3. Though he also advised his politicians to "become an authority in that area. Make sure that makes people start to follow you" (VVD, social media manager, 2013).
4. We did not interview people from 50Plus or the PVV.
5. In chapter 3 (Figure 3.1), we showed that candidates who tweet more tend to have more followers; so building a following really seems a function of lower-ranked politicians' social-media behavior.
6. Such as Maxime Verhagen (CDA), Ahmed Marcouch (PvdA), Martijn van Dam (PvdA), Diederik Samsom (PvdA), Ineke van Gent (GL), and Boris van der Ham (D66).
7. Regarding social media and privatization, see Kruikemeier (2014).
8. So far the few empirical studies have yielded mixed results: Larsson & Kalsnes, 2014; Luhiste & Sudulich, 2015. There are also some counterarguments: women in particular are said to be less tech savvy (see Luhiste & Sudulich, 2015; Van Zoonen, 2002).
9. Information on Facebook posts was not available.
10. In Norway there is no strong Green party, postmaterialist candidates are found most in the social democratic and the liberal party.
11. Our findings may apply to primaries, but given the winner-takes-all nature of the primary process, normalization is more likely in such a context.
12. For instance, see Goff (2013).
13. Based on the Dutch election survey, we can add that 97 percent of respondents indicate that the party is more important than the candidate in determining people's vote (see Jacobs & Spierings, 2014). Also, the potential of social media to show the personal side and private life of the politicians—privatization—is met very reluctantly by Dutch politicians, and Dutch public opinion does not seem thrilled either. See for instance joop.nl (2012).
14. In the case of the PVV, the list puller literally is the party and his account *is* the party's account.
15. It does bring them a bit closer to high-profile politicians, because they had no connection or public of their own before. However, in general they are still ranked lowly, and the list pullers are still way ahead of them.

6 Social Media Go "Glocal": The Local and European Arenas

1. In the 2014 local elections, they obtained one-third of all seats across the Netherlands.
2. We are very grateful for the collaboration with, and work of, Jeroen Hellebrekers and Willem van Sermondt (both Political Science Master's students at Radboud University).
3. The data on the number of followers were collected on May 15, 2014 for the four municipalities, which was after the elections and leads to slight overestimation of network sizes assuming that the networks are growing, particularly those of elected candidates. Most data were collected via Twitter's API's. If this was not possible, counting was done by hand.
4. On Facebook, the party account sometimes (though not always) performs better, which may indicate that the national party leadership lent a helping hand.
5. Stadspartij Groningen, Stadspartij Wageningen, Inwonerspartij toekomst Houten.
6. To be precise, 1 elderly party, 1 CU, 1 SGP, 1 SP, and 2 independent local parties did not have an account.
7. The position of Amsterdam falls between the local and national averages. The capital's list pullers are often well ahead of the rest of the candidates. The national attention they get might play an important role here.
8. Huub Bellemakers (who we interviewed), Peter Kwint, and Stijn Verbruggen are three of the ten candidates with more than 3,000 followers.
9. The only included parties are those that won seats or were represented in national politics. Based on data reported in Vergeer, Hermans, and Sams (2013).
10. This was calculated based on Vergeer, Hermans, and Sams (2013): all followers as also reported in Table 6.4 were added up and divided by 27—the number of candidates on Twitter—after detracting 1,600, which is a rough approximate of the number of followers of the party account (one-eighth of the 2015 network size).
11. Klout scores are influenced by the number of followers one has, but also the engagement one's messages trigger (see: Klout, 2015).
12. These differences were not statistically significant, but substantially they were (11.3% versus 14.9% and 6.7 versus 11.8 days before elections). Moreover, all candidates were included, so statistical significance has less value.
13. The odds ratio was 1.38 in 2009 and 1.26 in 2010: women had a 32 percent higher probability to be on Twitter than men in 2009, and in 2010 this was 16 percent. As these are raw numbers, the importance of party differences might explain this difference, as the postmaterialist parties fielded relatively more candidates than the other parties in the European elections than in the national elections. Also, given their second-order nature (as there is less at stake), the European lists might be more open to female candidates in general (see chapter 5).

14. A few studies do focus on the equalization-normalization debate in Web 1.0 online campaigns, such as Gibson, Lusoli, and Ward (2008).

7 Do Social Media Help Win Elections?

1. It is important to note that one should read the word "effect" as implying a correlation, not causation. After all, our analysis can only show a correlation after controlling for other explanations. It cannot directly prove causation, as this would require experimental research or process tracing (though the latter is less useful than the former when trying to explain voting behavior).

2. A similar argument holds for the list pullers of parties and the first women, therefore they have been excluded from the analyses here.

3. While tempting, such intraparty effects should not be translated to interparty effects. After all, a successful social-media candidate of, say, the progressive liberals may well have attracted extra votes which otherwise would have simply gone to another candidate of that same party.

4. The number of ethnic candidates in our dataset is extremely small and does not really allow for carrying out sensible statistical analyses.

5. Clearly voters can dislike the messages they receive. Nevertheless, an overall positive effect is to be expected, given the importance of (name) recognition (Grimmer, Messing, & Westwood, 2012). This is assumed to outweigh some tweets' possible negative "side effects" on some followers.

6. Facebook is different in this respect: it had more than 7.5 million users in 2012 (http://www.internetworldstats.com/stats9.htm). However, Dutch politicians mainly adopt the medium for private use.

7. Own calculations based on v410, v416, v430 v436; $n = 1,408$, weight: wgt1c; v416 and v436 refer to Internet and social networks sites together, but as e-mail (e.g., news letters) was asked about separately, actively being contacted via a website is very unlikely. The answers must thus mostly refer to social media; see Van der Kolk et al., 2013.

8. Based on own calculations using NKO data; 2012: $n = 1,677$, weight: wgt1c; 2010: $n = 2,15X$, weight: wgt4.

9. Our earlier work suggests that Hyves, the Dutch equivalent of Facebook until 2011, did not yield a significant effect (Spierings & Jacobs, 2014).

10. These usual suspects are: Incumbency (dummy); being well-known (dummy); media exposure, operationalized as the number of newspaper articles the year before the campaign, as well as by a second variable indicating the number of newspaper articles during the campaign period; whether or not the candidate had a personal website (dummy); whether or not the candidate had any previous campaign experience (dummy); her position on the list; whether or not the candidate occupied the last position on the list (dummy); was a woman (dummy); was the first ethnic nonwhite candidate on the list (dummy); was an ethnic nonwhite candidate on another position (dummy); and finally we included dummies for each of the parties (minus one—the reference category).

For an overview of the theoretical mechanisms behind each of these variables, see: Spierings and Jacobs, 2014.

11. The Conservative liberal Minister of Public Health, Edith Schippers (VVD) is a notable example of such a candidate. On May 24, 2015, she had 4,177 followers, though she had not yet posted a single tweet.

12. In 2010, the threshold turned out to be 15,694 preference votes.

13. Similar analyses for ethnic-minority or LGBT candidates would be interesting, but given the low number of such candidates, such analyses are unfeasible at the time.

14. The average 2012 female candidate had 518.5 followers and posted 2.41 messages a day. Plugging these values into the regression formula suggests that *ceteris paribus* such a candidate gained some 0.05 in preference-vote share.

15. When done properly, these studies show that social-media use is a severely biased predictor of election outcomes (e.g., Jungherr, Jürgens, & Schoen 2012; Gayo-Avello, 2012).

16. Sudulich and Wall (2010) find an effect in districts with an above median Internet penetration rate.

17. The deviant study is the one by Hansen and Kosiara-Pedersen. One possible explanation is that Hansen and Kosiara-Pedersen used a different operationalization of online communication. Their operationalization focused on the diversity of online media tools that a candidate claimed to use. Additionally they included an ordinal transformation of that variable for each candidate in a given district ("rank score"). Therefore this study is not fully comparable to the Dutch studies. It would be useful to see an analysis with an interaction effect between actual use and the actual number of followers rather than candidate survey data, as these cannot tap into the use and following of a candidate directly. Another potential explanation is that the imbalance might be due to publication bias, whereby non-significant findings are more difficult to publish. Finally it could be that Denmark is the exception to the rule. Ultimately, which of the three is correct is a matter of empirical assessment.

18. We suspect that this is due to the "I just don't buy it" argument often used by reviewers, which propels authors to levels of higher technical sophistication and additional control variables.

19. The only exception is the social-media manager of the ultraorthodox Christian party SGP, who considered Geert Wilders (PVV) and the Party of the Animals (PvdD) his biggest inspiration (Proos, 2014).

20. Though they might overestimate the interparty effect.

21. While the electoral effect for average candidates is modest, our analyses also suggest that candidates who tweet a lot can accumulate a more substantial vote bonus.

22. One important caveat is that several of these studies only examine the overall electoral effect of cyber campaigning in one election.

23. For instance, the Hansen and Kosiara-Pedersen (2014) study employed a survey question where the highest value indicated using social media daily. As we have seen, in the Netherlands one daily update would even be below

average in the early adoption phase. Furthermore such a cut-off point would not allow investigation into the diminished returns of additional posts. In the defense of the authors, often such measurements are the only available ones in the survey.

8   Conclusion

1. For a more elaborate definition, see Howard and Parks, 2012:359.
2. In chapter 2, we also suggested that populists—as parties for the people—might have the intrinsic motivation to use social media to connect to citizens but also noted that such parties have a strong motivation *not* to use social media as they are typically centrally organized and do not want to lose control. We therefore focus on the postmaterialist parties in this brief overview of the model.
3. One could even label this phase "unequalization" as the bigger parties actually strengthen their advantage over those smaller parties).
4. Especially if one considers that the other postmaterialists were clearly smaller than the bigger parties but were nonetheless on par with them.
5. In fact, the social-media manager of the CU did try: "We trained the politicians themselves, but they really did not have a clue, so you start from scratch. (...) We focused mainly on ten cities." (CU, social media manager, 2014).
6. Or to be precise: the SGP's youth wing gave the SGP politicians a Facebook page as a gift (SGP, social media manager, 2014).
7. At the local level, where the diffusion of social media is far slower than at the national and European level, the parties were still in the widespread-diffusion phase and voters have a lower demand for information via social media (cf. Klinger, Rösli, & Jarren, 2015).
8. The analysis of the electoral benefits of social media (chapter 7) also includes the early-adoption phase.

References

Agranoff, C., & Tabin, H. (2011). *Socially Elected: How To Win Elections Using Social Media*. West Jordan: Pendant Publishing.

Almond, G. A., Powell Jr., G. B., Strom, K., & Dalton, Russel J. (2003). *Comparative Politics Today: A World View*. Harlow: Longman.

Andeweg, R. B. (2005). The Netherlands: The sanctity of proportionality. In Gallagher, M. and Mitchell, P. (eds.), *The Politics of Electoral Systems*. Houndmills: Palgrave MacMillan.

Andeweg, R. B., & Irwin, G. A. (2005). *Governance and Politics of the Netherlands (Comparative Government and Politics)*. New York: Palgrave MacMillan.

André, A., Wauters, B., & Pilet, J. B. (2012). It's not only about lists: Explaining preference voting in Belgium. *Journal of Elections, Public Opinion & Parties*, *22*(3), 293–313.

Bailey, J., Steeves, V., Burkell, J., & Regan, P. (2013). Negotiating with gender stereotypes on social networking sites: From "bicycle face" to Facebook. *Journal of Communication Inquiry*, *37*(2), 91–112.

Bartlett, A. (2013). The 2013 Federal Election: The Greens campaign. In Johnson, C. and Wanna, J. (eds.) *Abbott's Gambit: the 2013 Australian Federal Election*, Canberra: ANU Press.

Beach, D., & Pedersen, R. B. (2013). *Process-tracing Methods: Foundations and Guidelines*. Ann Arbor: University of Michigan Press.

Bimber, B. (2012). Digital media and citizenship. *The SAGE Handbook of Political Communication*, 115–127.

Blumler, J. G., & Kavanagh, D. (1999). The third age of political communication: Influences and features. *Political communication*, *16*(3), 209–230.

Boix, C., & Stokes, S. C. (eds.). (2007). *The Oxford Handbook of Comparative Politics*. Oxford: Oxford University Press.

Bond, R. M., Fariss, C. J., Jones, J. J., Kramer, A. D., Marlow, C., Settle, J. E., & Fowler, J. H. (2012). A 61-million-person experiment in social influence and political mobilization. *Nature*, *489*(7415), 295–298.

Boogers, M. J. G. J. A. (2010). Het huis van de democratie na de gemeenteraadsverkiezingen: Achterstallig onderhoud? In Boogers, M., Meurs, E., Van Omme, A. M., Stein, M. F., & Vis, J. (eds.), *Jaarboek Vereniging van Griffiers*. Den Haag: SDU.

Bowler, S., Donovan, T., & van Heerde, J. 2005. The United States. In M. Gallagher, & P. Mitchell (eds.), *The Politics of Electoral Systems*. Oxford: Oxford University Press.

boyd, D., & Ellison, N. B. (2007). Social network sites: Definition, history, and scholarship. *Journal of Computer-Mediated Communication, 13*(1), 210–230.

Broersma, M., & Graham, T. (2012). Social media as beat: Tweets as a news source during the 2010 British and Dutch elections. *Journalism Practice, 6*(3), 403–419.

Castells, M. (1996). *The information age: Economy, society, and culture. Volume I: The rise of the network society.* Oxford: Blackwell.

CDA, social media manager EP (2014). Personal interview.

Celis, K., & Childs, S. (2008). Introduction: The descriptive and substantive representation of women: New directions. *Parliamentary Affairs, 61*(3), 419–425.

Celis, K., Childs, S., Kantola, J., & Krook, M. L. (2008). Rethinking women's substantive representation. *Representation, 44*(2), 99–110.

Celis, K., Childs, S., Kantola, J., & Krook, M. L. (2014). Constituting Women's Interests through Representative Claims. *Politics & Gender, 10*(02), 149–174.

Celis, K., & Erzeel, S. (2013). Gender and ethnicity: Intersectionality and the politics of group representation in the Low Countries. *Representation, 49*(4), 487–499.

Celis, K., & Spierings, N. (2014). Posting political descriptive underrepresentation away? An intersectional analysis of the use of social media by Dutch women and ethnic minorities candidates. Paper presented at Politicologenetmaal 2014, Maastricht.

Ceron, A., & d'Adda, G. (2015). E-campaigning on Twitter: The effectiveness of distributive promises and negative campaign in the 2013 Italian election. *New Media & Society,* 1461444815571915.

Chadwick, A. (2013). *The Hybrid Media System: Politics and Power.* Oxford: Oxford University Press.

Chen, P. J., & Smith, P. J. (2010). Adoption and use of digital media in election campaigns: Australia, Canada and New Zealand compared. *Public Communication Review, 1*(1), 3–26.

Chen, P. J. (2015). New Media in the Electoral Context: The new normal. In Johnson, C. and Wanna, J.(eds.). *Abbott's Gambit: The 2013 Australian Federal Election,* Canberra: ANU Press.

Childs, S., & Lovenduski, J. (2013). Political representation. In Waylen, G., Celis, K., Kantola J., & Weldon, L. (eds.), *The Oxford Handbook on Gender and Politics.* New York: Oxford University Press.

Colomer, J. M. (2011). *Personal Representation. The Neglected Dimension of Electoral Systems.* Colchester: ECPR Press.

Conway, B. A., Kenski, K., & Wang, D. (2013). Twitter use by presidential primary candidates during the 2012 campaign. *American Behavioral Scientist, 57*(11), 1596–1610.

Cornfield, M. (2005). *The Internet and Campaign 2004: A Look Back at the Campaigners.* Washington, DC: Pew Internet & American Life Project. Retrieved March 10, 2015 from http://www.pewinternet.org/pdfs/Cornfield_commentary.pdf.

Crawford, K. (2009). Following you: Disciplines of listening in social media. *Continuum: Journal of Media & Cultural Studies, 23*(4), 525–535.

CU, social media manager (2014). Personal interview.

Curran, J., Coen, S., Soroka, S., Aalberg, T., Hayashi, K., Hichy, Z.,…Tiffen, R. (2014). Reconsidering "virtuous circle" and "media malaise" theories of the media: An 11-nation study. *Journalism, 15*(7), 815–833.

D66, MP (2013). Telephone interview.

D66, social media webcare (2014). Personal interview.

D'Alessio, D. (1997). Use of the World Wide Web in the 1996 US election. *Electoral Studies, 16*(4), 489–500.

De Valk, E. Timmermans terug op Facebook—niet voor "vrienden," maar voor "fans." http://www.nrc.nl/nieuws/2012/12/19/timmermans-terug-op-facebook-geen-vrienden-maar-fans/. Accessed August 13, 2013.

De Vreese, C. H. (2001). Europe'in the news a cross-national comparative study of the news coverage of key EU events. *European Union Politics, 2*(3), 283–307.

De Vries, J. (2012). Strijden voor een zetel: "Jammer dat ik mijn campagne-Cadillac moet inleveren." http://www.volkskrant.nl/dossier-vk-dossier-verkiezingen-van-2012/strijden-voor-een-zetel-jammer-dat-ik-mijn-campagne-cadillac-moet-inleveren-a3314515/. Accessed February 13, 2014.

Dijkstra, J.-K. (2012). Verkiezingsstrijd via social media: verschil in aanpak VS & Nederland. http://www.frankwatching.com/archive/2012/08/23/verkiezingsstrijd-via-social-media-verschil-in-aanpak-vs-en-nederland/20150417/. Accessed February 13, 2014.

Dolezal, M. (2015). Online campaigning by Austrian political candidates: Determinants of using personal Websites, Facebook, and Twitter. *Policy & Internet, 7*(1), 103–119.

Dugan, Lauran (2012). "The US Has The Most Twitter Users, But The Netherlands Is More Active [STATS]." All Twitter, February 1, 2012. http://www.mediabistro.com/alltwitter/the-us-has-the-most-twitter-users-but-thenetherlands-is-more-active-stats_b18172. Accessed February 13, 2015.

Enyedi, Z. (2008). The social and attitudinal basis of political parties: Cleavage politics revisited. *European Review, 16*(3), 287–304.

Europees Parlement Informatiebureau Nederland (2014). De Europese verkiezingen van 22 tot 25 mei: debatteer mee op Twitter en op Facebook! #EP2014. http://www.europeesparlement.nl/nl/press_release/pr-2014/2014-april/pr-2014-april22.html. Accessed February 2, 2015.

Evans, H. K., Cordova, V., & Sipole, S. (2014). Twitter style: An analysis of how house candidates used Twitter in their 2012 campaigns. *PS: Political Science & Politics, 47*(2), 454–462.

Farrell, D. (2011). *Electoral Systems. A Comparative Introduction*. Houndmills: Palgrave Macmillan.

Gallagher, M. (2015). Election Indices. https://www.tcd.ie/Political_Science/staff/michael_gallagher/ElSystems/Docts/ElectionIndices.pdf. Accessed May 5, 2015.

Gallagher, M., & Mitchell, P. (eds.). (2005). *The Politics of Electoral Systems*. Oxford: Oxford University Press.

Gayo-Avello, D. (2012). "I Wanted to Predict Elections with Twitter and all I got was this Lousy Paper"—A Balanced Survey on Election Prediction using Twitter Data. *arXiv preprint arXiv:1204.6441*.

Gerring, J. (2007). *Case Study Research: Principles and Practices*. Cambridge: Cambridge University Press.

Gibson, R. K. (2009). New media and the revitalisation of politics. *Representation*, 45(3), 289–299.

Gibson, R. K., Lusoli, W., & Ward, S. (2008). Nationalizing and normalizing the local? A comparative analysis of online candidate campaigning in Australia and Britain. *Journal of Information Technology & Politics*, 4(4), 15–30.

Gibson, R. K., & McAllister, I. (2006). Does cyber-campaigning win votes? Online communication in the 2004 Australian election. *Journal of Elections, Public Opinion and Parties*, 16(3), 243–263.

Gibson, R. K., & McAllister, I. (2011). Do online election campaigns win votes? The 2007 Australian "YouTube" election. *Political Communication*, 28(2),2 27–244.

Gibson, R. K., & McAllister, I. (2014). Normalising or equalising party competition? Assessing the impact of the web on election campaigning. *Political Studies*. doi:10.1111/1467–9248. 63(3), 529–547.

Gibson, R., & Römmele, A. (2001). Changing campaign communications: A party-centered theory of professionalized campaigning. *The Harvard International Journal of Press/Politics*, 6(4), 31–43.

Gibson, R., Römmele, A., & Williamson, A. (2014). Chasing the digital wave: International perspectives on the growth of online campaigning. *Journal of Information Technology & Politics*, 11(2), 123–129.

Gibson, R., & Ward, S. (2000). A proposed methodology for studying the function and effectiveness of party and candidate web sites. *Social Science Computer Review*, 18(3), 301–319.

Gibson, R. K., & Ward, S. (2012). Political organizations and campaigning online. *The SAGE Handbook of Communication*, 62–74.

Gil de Zúñiga, H., Jung, N., & Valenzuela, S. (2012). Social media use for news and individuals' social capital, civic engagement and political participation. *Journal of Computer-Mediated Communication*, 17(3), 319–336.

GL, social media manager (2013). Personal interview.

Goertz, G. (2006). *Social Science Concepts: A User's Guide*. Princeton, NJ: Princeton University Press.

Goff, K. (2013). Single presidents of the U.S. Could one be elected? http://www.theroot.com/articles/politics/2013/01/single_presidents_of_the_us_could_one_be_elected_today.4.html. Accessed February 1, 2015.

Golbeck, J., Grimes, J. M., & Rogers, A. (2010). Twitter use by the US Congress. *Journal of the American Society for Information Science and Technology*, *61*(8), 1612–1621.

Goodman, E. (2009) Journalism in the Twitter Era. http://www.truthdig.com /report/item/20090624_journalism_in_a_twitter_era. Accessed December 1, 2014.

Graber, D. A., & Smith, J. M. (2005). Political communication faces the 21st century. *Journal of Communication*, *55*(3), 479–507.

Graham, T., Jackson, D., & Broersma, M. (2014). New platform, old habits? Candidates' use of Twitter during the 2010 British and Dutch general election campaigns. *New Media & Society*, 1461444814546728.

Grimmer, J., Messing, S., & Westwood, S. J. (2012). How words and money cultivate a personal vote: The effect of legislator credit claiming on constituent credit allocation. *American Political Science Review*, *106*(04), 703–719.

Hallin, D. C., & Mancini, P. (2004). *Comparing Media Systems: Three Models of Media and Politics*. Cambridge: Cambridge University Press.

Hansen, K. M., & Kosiara-Pedersen, K. (2014). Cyber-campaigning in Denmark: Application and effects of candidate campaigning. *Journal of Information Technology & Politics*, *11*(2), 206–219.

Harcup, T., & O'neill, D. (2001). What is news? Galtung and Ruge revisited. *Journalism Studies*, *2*(2), 261–280.

Harlow, S., & Harp, D. (2012). Collective action on the Web: A cross-cultural study of social networking sites and online and offline activism in the United States and Latin America. *Information, Communication & Society*, *15*(2), 196–216.

Herbst, S. (2011). Un(Numbered) Voices? Reconsidering the meaning of public opinion in a digital age. In Goidel, Robert K. (ed.), *Political Polling in the Digital Age*. Baton Rouge, LA: Louisiana State University Press.

Howard, P. N., & Parks, M. R. (2012). Social media and political change: Capacity, constraint, and consequence. *Journal of Communication*, *62*(2), 359–362.

Inglehart, R. (1997). *Modernization and Postmodernization: Cultural, Economic, and Political Change in 43 Societies* (Vol. 19). Princeton, NJ: Princeton University Press.

Inwonerspartij Toekomst Houten (2014). Telephone interview.

Jackson, N. A., & Lilleker, D. G. (2009). Building an architecture of participation? Political parties and Web 2.0 in Britain. *Journal of Information Technology & Politics*, *6*(3–4), 232–250.

Jackson, N., & Lilleker, D. (2011). Microblogging, constituency service and impression management: UK MPs and the use of Twitter. *The Journal of Legislative Studies*, *17*(1), 86–105.

Jacobs, K. (2011). *The Power or the People? Direct Democratic and Electoral Reforms in Austria, Belgium and the Netherlands*. PhD dissertation. Radboud Universiteit Nijmegen.

Jacobs, K., & Leyenaar, M. (2011). A conceptual framework for major, minor, and technical electoral reform. *West European Politics*, *34*(3), 495–513.

Jacobs, K., & Spierings, N. (2014)....Maar win je er stemmen mee? De impact van Twittergebruik door politici bij de Nederlandse Tweede Kamerverkiezingen van 12 september 2012. *Tijdschrift Voor Communicatiewetenschap, 42*(1), 22–38.

Jacobs, K., & Spierings, N. (2015). De impact van digitale campagnemiddelen op de personalisering van politieke partijen in Nederland (2010–2014). *Res Publica, 57*(1), 57–77.

Jacobs, K., Sudulich, L., Gaglio, M., Pilet, J.-B., & Boireaau, M. (2015). Something Old or Something New? Assessing the Impact of Traditional versus Social Media on Preference Voting. Paper presented at the ECPR Joint Sessions of Workshops 2015, Warsaw.

Joop.nl (2012). Diederik Samsom zeer persoonlijk in verkiezingsspot. http://www .joop.nl/politiek/detail/artikel/14743_diederik_samsom_zeer_persoonlijk_in _verkiezingsspot. Accessed March 14, 2015.

Jungherr, A. (2014). Twitter in politics: A comprehensive literature review. *Available at SSRN 2402443.*

Jungherr, A., Jürgens, P., & Schoen, H. (2012). Why the pirate party won the german election of 2009 or the trouble with predictions: A response to tumasjan, a., sprenger, to, sander, pg, & welpe, im "predicting elections with twitter: What 140 characters reveal about political sentiment." *Social Science Computer Review, 30*(2), 229–234.

Karlsen, R. (2011). Still broadcasting the campaign: On the Internet and the fragmentation of political communication with evidence from Norwegian electoral politics. *Journal of Information Technology & Politics, 8*(2), 146–162.

Karvonen, L. (2010). *The Personalisation of Politics: A Study of Parliamentary Democracies.* Coulchester: ECPR Press.

Katz, J. E., Barris, M., & Jain, A. (2013). *The Social Media President: Barack Obama and the Politics of Digital Engagement.* New York: Palgrave Macmillan.

Klinger, U., Rösli, S., & Jarren, O. (2015). To Implement or Not to Implement? Participatory Online Communication in Swiss Cities. *International Journal of Communication, 9,* 1926–1946.

Klompenhouwer, L. (2014). PvdA-Kamerleden uit fractie na kritiek op Asscher. http://www.nrc.nl/nieuws/2014/11/13/pvda-kamerleden-uit-fractie-na-kritiek -op-asscher/. Accessed February 26, 2015.

Klout (2015) The Klout score. https://klout.com/corp/score. Accessed May 23, 2015.

Koc-Michalska, K., Gibson, R., & Vedel, T. (2014). Online Campaigning in France, 2007–2012: Political Actors and Citizens in the Aftermath of the Web. 2.0 Evolution. *Journal of Information Technology & Politics, 11*(2), 220–244.

Koc-Michalska, K., Lilleker, D. G., Surowiec, P., & Baranowski, P. (2014). Poland's 2011 online election campaign: new tools, new professionalism, new ways to win votes. *Journal of Information Technology & Politics, 11*(2), 186–205.

Kriesi, H. (2012). Personalization of national election campaigns. *Party Politics, 18*(6), 825–844.

Kruikemeier, S. (2014). How political candidates use Twitter and the impact on votes. *Computers in Human Behavior, 34,* 131–139.

Kruikemeier, S., van Noort, G., Vliegenthart, R., & de Vreese, C. H. (2013). Getting closer: The effects of personalized and interactive online political communication. *European Journal of Communication*, 0267323112464837.

Kruikemeier, S., van Noort, G., Vliegenthart, R., & de Vreese, C. H. (2014). Unraveling the effects of active and passive forms of political Internet use: Does it affect citizens' political involvement?. *New Media & Society*, *16*(6), 903–920.

Larsson, A. O. (2013). "Rejected bits of program code": Why notions of "Politics 2.0" remain (mostly) unfulfilled. *Journal of Information Technology & Politics*, *10*(1), 72–85.

Larsson, A. O., & Kalsnes, B. (2014). 'Of course we are on Facebook': Use and non-use of social media among Swedish and Norwegian politicians. *European Journal of Communication*, 0267323114531383.

Larsson, A. O., & Svensson, J. (2014). Politicians online—Identifying current research opportunities. *First Monday*, *19*(4).

Lasorsa, D. L., Lewis, S. C., & Holton, A. E. (2012). Normalizing Twitter: Journalism practice in an emerging communication space. *Journalism Studies*, *13*(1), 19–36.

Lassen, D. S., & Brown, A. R. (2011). Twitter: The electoral connection? *Computer Review*, *29*, 419–436. doi:10.1177/0894439310382749.

Lee, F. L. (2012). The life cycle of iconic sound bites: Politicians' transgressive utterances in media discourses. *Media, Culture & Society*, *34*(3), 343–358.

Leijenaar, M. (1997). *How to Create a Gender Balance in Political Decision-Making: A Guide to Implementing Policies for Increasing the Participation of Women in Political Decision-Making*. Luxembourg: Office for Official Publications of the European Communities.

Lenski, G. (2005). *Ecological-Evolutionary Theory: Principles and Applications*. Colorado: Paradigm.

Lijphart, A. (1999). *Patterns of Democracy: Government Forms and Performance in Thirty-Six Democracies*. New Haven, CT: Yale University Press.

Lilleker, D. G., & Jackson, N. A. (2010). Towards a more participatory style of election campaigning: The impact of Web 2.0 on the UK 2010 general election. *Policy & Internet*, *2*(3), 69–98.

Luhiste, M., & Sudulich, L. (2015). Gender Differences in the Use of New Technologies: Candidates' campaigning in comparative perspective. Paper presented at the ECPR Joint Sessions of Workshops 2015, Warsaw.

Lutz, G. (2010). First come, first served: The effect of ballot position on electoral success in open ballot PR elections. *Representation*, *46*(2), 167–181.

Margolis, M., & Resnick, D. (2000). *Politics as Usual: The Cyberspace "Revolution"*. Thousand Oaks, CA: Sage Publications.

Margolis, M., Resnick, D., & Wolfe, J. D. (1999). Party competition on the Internet in the United States and Britain. *The Harvard International Journal of Press/Politics*, *4*(4), 24–47.

Martin, A. (2013). Politicians struggle with authenticity on Twitter, self-censorship prevails. http://www.wired.co.uk/news/archive/2013-06/14/politicians-twitter. Accessed February 1, 2015.

Mascheroni, G., & Mattoni, A. (2013). Electoral Campaigning 2.0—The Case of Italian Regional Elections. *Journal of Information Technology & Politics, 10*(2), 223–240.

McAllister, I. (2007). The personalization of politics. *The Oxford Handbook of Political Behavior.* Oxford: Oxford University Press, 571–588.

McCall, L. (2005). The Complexity of Intersectionality. *SIGNS: Journal of Women in Culture and Society, 30*(31), 1771–1800.

Miniwatts Marketing Group (2015). Internet World Stats. http://www.internet worldstats.com. Accessed March 23, 2015.

Mudde, C. (2004). The populist zeitgeist. *Government and opposition, 39*(4), 542–563.

Mudde, C. (2007). *Populist Radical Right Parties in Europe* (Vol. 22, No. 8). Cambridge: Cambridge University Press.

Mügge, L., & De Jong, S. (2013). Intersectionalizing European politics: Bridging gender and ethnicity. *Politics, Groups, and Identities, 1*(3), 380–389.

Nash, J. C. (2008). Re-thinking intersectionality. *Feminist Review, 89*(1), 1–15.

Negroponte, N. (1995). *Being Digital.* New York: Knopf.

NKO (2012). Nationaal Kiezersonderzoek.

Norris, P. (2000). *A virtuous circle: Political communications in postindustrial societies.* Cambridge: Cambridge University Press.

Norris, P., Curtice, J., Sanders, D., Scammell, M., & Semetko, H. A. (1999). *On Message: Communicating the Campaign.* London: SAGE.

NOS (2012). 82% twitter-volgers Buma nep. http://nos.nl/artikel/399093-82 -twittervolgers-buma-nep.html. Accessed August 13, 2013.

NU.nl (2012). Hernandez en Kortenoeven stappen uit PVV-fractie. http://www .nu.nl/politiek/2850048/hernandez-en-kortenoeven-stappen-pvv-fractie.html. Accessed April 30, 2015.

NU.nl (2014). Kuzu en Öztürk starten partij Denk. http://www.nu.nl/politiek /3989094/kuzu-en-ozturk-starten-partij-denk.html. Accessed February 26, 2015.

Oosterveer, D. (2012). Social media 2012 in cijfers. Hoe staan de sociale media er eind 2012 voor? http://www.marketingfacts.nl/berichten/social-media-2012-in -cijfers. Accessed August 12, 2013.

Oosterveer, D. (2013). Social media in Nederland 2013: Groei van gebruik Twitter en Facebook afgevlakt. http://www.marketingfacts.nl/berichten/social-media-in -nederland-twitter-enfacebook-het-meest-actief-gebruikt. Accessed August 13, 2013.

Oosterveer, D. (2014). De laatste cijfers van het socialmediagebruik in Nederland. Alle cijfers op een rijtje van o.a. Twitter, Facebook, LinkedIn, Google+, Pinterest, Instagram en meer. http://www.marketingfacts.nl/berichten/socialmediagebruik -in-nederland-update-maart-2014. Accessed February 5, 2014.

Parmelee, J. H., & Bichard, S. L. (2011). *Politics and the Twitter Revolution: How Tweets Influence the Relationship between Political Leaders and the Public.* Lanham, MD: Lexington Books.

Peterson, R. D. (2012). To tweet or not to tweet: Exploring the determinants of early adoption of Twitter by House members in the 111th Congress. *The Social Science Journal, 49*(4), 430–438.

Plasser, F., & Plasser, G. (2002). *Global Political Campaigning: A Worldwide Analysis of Campaign Professionals and Their Practices.* Westport, CT: Greenwood Publishing Group.

Poguntke, T., & Webb, P. (2005). *The Presidentialization of Politics: A Comparative Study of Modern Democracies.* Oxford: Oxford University Press.

Polat, R. K. (2005). The Internet and political participation exploring the explanatory links. *European journal of communication, 20*(4), 435–459.

Pollard, C. (2013) Election showcases power of social media. http://www.theage .com.au/smallbusiness/smallbiz-marketing/election-showcases-power-of-social -media-20130815-2rxkn.html. Accessed August 12, 2013.

PvdA, MP (2013). Personal interview.

PvdA, social media consultant (2014). Personal interview.

PvdA, social media manager (2013). Personal interview.

PvdD, social media manager (2014). Personal interview.

Quintelier, E., & Theocharis, Y. (2012). Online political engagement, Facebook, and personality traits. *Social Science Computer Review*, 31(3), 280–290.

Reif, K., & Schmitt, H. (1980). Nine Second-order National Elections–a Conceptual Framework for the Analysis of European Election Results. *European journal of Political Research*, 8(1), 3–44.

Rogers, E. M. (2002). Diffusion of preventive innovations. *Addictive Behaviors, 27*(6), 989–993.

Ruedin, D. (2013). *Why Aren't They There? The Political Representation of Women, Ethnic Groups and Issue Positions in Legislatures.* Colchester: ECPR Press.

Saward, M. (2010). *The Representative Claim.* Oxford: Oxford University Press.

Scarrow, S. E. (2007). Political finance in comparative perspective. *Annual Review of Political Science, 10*, 193–210.

Schectman, J. (2009). Iran's twitter revolution? maybe not yet. *Business Week, 17*(07).

Schlozman, K., Verba, S. & Brady, H. (2012). *The Unheavenly Chorus: Unequal Political Voice and the Broken Promise of American Democracy.* Princeton, NJ: Princeton University Press.

Schumacher, G., & Rooduijn, M. (2013). Sympathy for the "devil"? Voting for populists in the 2006 and 2010 Dutch general elections. *Electoral Studies, 32*(1), 124–133.

Schweitzer, E. J. (2011). Normalization 2.0: A longitudinal analysis of German online campaigns in the national elections 2002–9. *European Journal of Communication*, 26(4), 310–327.

SGP, social media manager (2014). Personal interview.

Shirky, C. (2011). The political power of social media: Technology, the public sphere, and political change. *Foreign affairs*, 90(1), 28–41.

Short, J., Williams, E., & Christie, B. (1976). *The Social Psychology of Telecommunications.* London: Wiley.

Small, T. A. (2008). Equal access, unequal success—major and minor Canadian parties on the net. *Party Politics, 14*(1), 51–70.

Small, T. A. (2011). What the hashtag? A content analysis of Canadian politics on Twitter. *Information, Communication & Society, 14*(6), 872–895.

Smith, S. (2012). #TK2012: welke partij wint op Facebook? (update 10/9). http://www.likeconomics.nl/2012/09/tk2012-welke-partij-wint-op-facebook/. Accessed August 13, 2013.

Southern, R. (2015). Is Web 2.0 Providing a Voice for Outsiders? A Comparison of Personal Web Site and Social Media Use by Candidates at the 2010 UK General Election. *Journal of Information Technology & Politics* (ahead-of-print), 1–17.

SP, campaign leader/senator (2013). Personal interview.

SP, social media manager (2013). Personal interview.

Spierings, N. (2012). The inclusion of quantitative techniques and diversity in the mainstream of feminist research. *European Journal of Women's Studies, 19*(3), 331–347.

Spierings, N., & Jacobs, K. (2014). Getting personal? The impact of social media on preferential voting. *Political Behavior, 36*(1), 215–234.

Spierings, N., Zaslove, A., Mügge, L. M., & de Lange, S. L. (2015). Gender and populist radical-right politics: an introduction. *Patterns of Prejudice, 49*(1–2), 3–15.

Steinfield, C., Ellison, N. B., & Lampe, C. (2008). Social capital, self-esteem, and use of online social network sites: A longitudinal analysis. *Journal of Applied Developmental Psychology, 29*(6), 434–445.

Stromer-Galley, J. (2004). Interactivity-as-product and interactivity-as-process. *The Information Society, 20*(5), 391–394.

Sudulich, M. L., & Wall, M. (2010). "Every little helps": Cyber-campaigning in the 2007 Irish general election. *Journal of Information Technology & Politics, 7*(4), 340–355.

Suiter, J. (2015). Political Campaigns and Social Media: A Study of# mhe13 in Ireland. *Irish Political Studies, 30*(2), 299–309.

Sundar, S. S., Kalyanaraman, S., & Brown, J. (2003). Explicating Web Site interactivity impression formation effects in political campaign sites. *Communication Research, 30*(1), 30–59.

Swigger, N. (2013). The Online Citizen: Is Social Media Changing Citizens' Beliefs About Democratic Values?. *Political Behavior, 35*(3), 589–603.

Tedesco, J. C. (2008). Changing the channel: Use of the internet for communicating about politics. In Kaid, L. L. (ed.), *Handbook of Political Communication Reserach* (pp. 507–528). New York: Routledge.

Thijssen, P., & Jacobs, K. (2004). Determinanten van voorkeurstemproporties bij (Sub-)locale Verkiezingen. De Antwerpse districtsraadsverkiezingen van 8 oktober 2000. *Res Publica, 46*(4), 460–485.

Utz, S. (2009). The (potential) benefits of campaigning via social network sites. *Journal of Computer-Mediated Communication, 14*(2), 221–243.

Vaccari, C. (2008). Surfing to the Elysee: The Internet in the 2007 French elections. *French Politics*, *6*(1), 1–22.

Valenzuela, S., Park, N., & Kee, K. F. (2009). Is there social capital in a social network site?: Facebook use and college students' life satisfaction, trust, and participation1. *Journal of Computer-Mediated Communication*, *14*(4), 875–901.

Van Aelst, P., Sheafer, T., & Stanyer, J. (2012). The Personalization of Mediated Political Communication: A Review of Concepts, Operationalizations and Key Findings, *Journalism*, *13*(2), 203–220.

Van Biezen, I. (2004). Political parties as public utilities. *Party Politics*, *10*(6), 701–722.

Van der Kolk, H., Tillie, J., Van Erkel, P., Van der Velden, M., & Damstra, A. (2013). *Dutch Parliamentary Election Study 2012*. Amsterdam: DANS, CBS, SKON.

Van Evera, S. (1997). *Guide to Methods for Students of Political Science*. New York: Cornell University Press.

Van Holsteyn, J. J., & Andeweg, R. B. (2012). Tweede Orde Personalisering: Voorkeurstemmen in Nederland. *Res Publica*, *54*(2), 163–91.

Van Kessel, S. (2015). *Populist Parties in Europe: Agents of Discontent?*. Houndsmills: Palgrave Macmillan.

Van Zoonen, L. (2002). Gendering the Internet claims, controversies and cultures. *European Journal of Communication*, *17*(1), 5–23.

Vergeer, M., & Hermans, L. (2013). Campaigning on Twitter: Microblogging and online social networking as campaign tools in the 2010 general elections in the Netherlands. *Journal of Computer-Mediated Communication*, *18*(4), 399–419.

Vergeer, M., Hermans, L., & Sams, S. (2011). Is the voter only a tweet away? Micro blogging during the 2009 European Parliament election campaign in the Netherlands. *First Monday*, *16*(8).

Vergeer, M., Hermans, L., & Sams, S. (2013). Online social networks and micro-blogging in political campaigning The exploration of a new campaign tool and a new campaign style. *Party Politics*, *19*(3), 477–501.

Vliegenthart, R. (2012). The professionalization of political communication? A longitudinal analysis of Dutch election campaign posters. *American Behavioral Scientist*, *56*(2), 135–150.

Wagner, K. M., & Gainous, J. (2009). Electronic grassroots: Does online campaigning work?. *The Journal of Legislative Studies*, *15*(4), 502–520.

Wauters, B., Weekers, K., & Maddens, B. (2010). Explaining the number of preferential votes for women in an open-list PR system: an investigation of the 2003 federal elections in Flanders (Belgium). *Acta Politica*, *45*(4), 468–490.

Weber Shandwick (2014). Twitter en de Tweede Kamer. http://webershandwick.nl/wp-content/uploads/2014/03/Twitter-en-de-Tweede-Kamer.-Weber-Shandwick.pdf. Accessed February 24, 2015.

Weldon, S. L. (2002). Beyond bodies: Institutional sources of representation for women in democratic policymaking. *The Journal of Politics*, *64*(04), 1153–1174.

Wilson, J. 2009. Can Social Media Save Iran? New Matilda (5th November), http://newmatilda.com/2009/11/05/can-social-media-save-iran

Wolfsfeld, G., Segev, E., & Sheafer, T. (2013). Social media and the Arab Spring politics comes first. *The International Journal of Press/Politics*, *18*(2), 115–137.

Woollaston, Victoria (2013). 'The meteoric rise of social networking in the UK: Britons are the second most prolific Facebook and Twitter users in EUROPE with a fifth of over 65s now using these sites' *Daily Mail*. June 13, 2013. http://www.dailymail.co.uk/sciencetech/article-2340893/Britons-second-prolific-Facebook-Twitter-users-EUROPE-fifth-aged-65.html.

Zantingh, P. (2012). Hero Brinkman stapt uit de PVV-fractie, maar blijft gedogen. http://www.nrc.nl/nieuws/2012/03/20/hero-brinkman-stapt-uit-de-pvv-fractie/. Accessed April 30, 2015.

Zhang, W., Johnson, T. J., Seltzer, T., & Bichard, S. (2010). The revolution will be networked: The influence of social network sites on political attitudes and behaviors. *Social Science Computer Review*, 28, 75–92.

Index

GPSR Compliance
The European Union's (EU) General Product Safety Regulation (GPSR) is a set
of rules that requires consumer products to be safe and our obligations to
ensure this.

If you have any concerns about our products, you can contact us on

ProductSafety@springernature.com

In case Publisher is established outside the EU, the EU authorized
representative is:

Springer Nature Customer Service Center GmbH
Europaplatz 3
69115 Heidelberg, Germany

www.ingramcontent.com/pod-product-compliance
Lightning Source LLC
Chambersburg PA
CBHW070942050326
40689CB00014B/3309